Galileo and its Applications

Tools for the Study of Cognitive and Cultural Processes

Joseph Woelfel

University at Buffalo

State University of New York

The Galileo System

RAH Press

Buffalo, NY

Joseph Woelfel/RAH Press

ISBN – 13 1530687985

ISBN – 10 1530687985

Print Edition, 2018

We are grateful to *Communication & Science Journal* for
permission to reprint some material previously published in
the journal.

Rah Press

"It's always fun to have your models validated, but it is way more fun to have them trashed," lead author Randy Gladstone of the Southwest Research Institute told the Los Angeles Times. "Finding out you are completely wrong is a great part of science."

FIGURES

Preface

Human knowledge springs from two and only two sources: direct personal observation (what George Herbert Mead called self-reflective acts) and the influence of others (what Haller, et. al. called "significant other influence.") When a culture has a poorly developed observational system, as in a prescientific culture that does not yet have agreed upon measurement standards, the Hindu number system and the like, direct observation is inhibited, and human attitudes and beliefs are dominated by the opinions of others. In this context, opinions wash over the culture, unchecked by confrontations with direct sensory experience, until they run their course, then fade and are replaced by the "next big thing."

While business analysts certainly have sophisticated means to determine the value of their bank accounts, their observational system for learning about people's attitudes and beliefs is pretty much restricted to categories and five (or fewer) point scales. (Can you imagine a business estimating its earnings and expenses on five-point scales? In the social sciences, although the same mechanisms for measuring monetary value also exist, they are seldom used. Here, for example, is a standard item for measuring a respondent's income from a commercial online survey supplier:

What is your current income in dollars?

Under $10,000

$10,000-$19,999

$20,000-$29,999

$30,000-$39,999

$40,000-$49,999

$50,000-$74,999

$75,000-$99,999

$100,000-$150,000

Over $150,000

On this scale, the principle at your local high school, a top full professor at an R1 University, Jeff Bezos and the Koch Brothers earn about the same.

Because of this cultural nearsightedness, both the business research community and the social science community have experienced the rise and fall of these Durkheimian currents of opinion. In The social sciences, *positivism* morphed into the *Test Against the Null Hypothesis (TANH) model,* which coexisted for a while with *verstehen* and *gestalt,* then *symbolic interactionism, existential phenomenology* and *postmodernism,* among many others. Pockets of each still exist, and, although Likert-type and other categorical scales and statistical significance levels still dominate the quantitative school, attacks on the TANH model are rising quickly. Crude measurements have left the social science community too blind to tell whether one theory is better or worse than another; after nearly a century of research, *not one theory* has been decisively rejected.

The business community has experienced similar trends, like *best practices, six sigma, business process reengineering, matrix management, management by consensus, core competence, management by objectives* and more. As this is written, the business community is currently wending its way through a crescendo of *analytics,* which is inspired by a wealth of "big data." Somehow the hope has arisen that, with so much data available, clever "artificially intelligent" analytics can sort through the horrendous mix of poorly measured, atheoretical information that gluts the Internet and discover some hidden pearls that will supercharge their businesses.

As Einstein taught us, it is actually theory that determines what we can observe, and the notion that superior analytics will enable us to find treasures in trash data without theoretical guidance represents once again the triumph of hope over experience. Perhaps, by the time you read this, the futility of hoping the massive sifting and winnowing of junk data will produce the hidden jewels we desire will have made itself apparent, perhaps not. But there is a better way.

11

Apology

This book tells the story of The Galileo System and its applications, a rational alternative to the current model of social science research. The first published Galileo study – Gail Wisan's dissertation research at the University of Illinois – was done 45 years before these words are being written[1]. Computer graphics were nonexistent. Computers could not write lower-case letters. Software was written on bricks of Hollerith cards, and handed though a "dispatch window" to run; results *might* be ready in two or three days. Over the ensuing 45 years, graphical display has improved dramatically. In this volume, work spanning nearly half a century is discussed, and the temptation to replace older graphics with current renderings is strong. Some of the older plots and graphics are crude and hard to read.

Even some of the modern graphical materials are optimized for display on high definition color monitors. These are nearly illegible when presented three inches wide in black and white print format. Nonetheless, in the interests of historical authenticity, we've not replaced older graphic contents or redone the video graphics for print format. All the graphic figures in this book are available on the Galileo Website so you can view them on your own high-resolution screen.

Chapter 1:

The Sociology of Sociology:
The unexamined discipline is not worth pursuing

This section is titled "The Sociology of Sociology" because I am myself a sociologist by training, but I believe that it applies as well to the social sciences in general. I've said elsewhere[2] that the social sciences constitute a culture, and it's fair to say that each of the various social sciences are subcultures within that general culture. Like any culture they share certain fundamental beliefs and values, norms, folkways and mores, collective representations and social facts. Like any real as opposed to fictional ideal culture, they also contain cleavages, disagreements, cliques, social networks and standing, irresolvable disputes. Within all this disparity, there exists a persistent subculture of social scientists who consider themselves scientists.

Not all social scientists believe social science is or should be science. Some social scientists chafe at the nomenclature because they believe science does not apply to human beings. You can find them on academic committees that argue about whether we should be called social science or social studies. This book is concerned only with those who believe what they do is science.

Social thought is very old, but social science is relatively new. Early sociologists like August Comte, impressed by the successes of 18th Century physical science, insisted that sociology, too, should be a science, and he even went so far as to claim sociology to be Queen of the Sciences. Most early sociologists, like Comte himself, had no background in science whatsoever. Of course, there are some exceptions, like Wilhelm Wundt, who studied medicine at Heidelberg and served as an assistant to the brilliant scientist Hermann Helmholtz. Wundt established the first psychology laboratory dedicated to the scientific study of psychology, and shortly thereafter, Emile Durkheim, who studied briefly with Wundt, laid down rules for the scientific study of society in his *Rules of the Sociological Method* [3]. For the most part, however, early social scientists were devoid of scientific backgrounds [2].

The real impetus for the rise of social *science* came with the formation of the National Science Foundation in the United States after World War II. Initially, the NSF proposed to fund only research from the physical sciences, but strong political pressure from social scientists and university administrators around the country succeeded in adding a small social science component. Committees were formed and textbooks written and published, such as *Research Methods in Social Relations*[4], which was revised and republished repeatedly for decades and used across most social science disciplines [5,6]. These formed – and continue to form – the core of what social scientists believe to be the scientific method.

There were two fundamental problems underlying the rapid establishment of social science's scientific methodology: first, the social scientists themselves, including the textbook authors, had little if any actual acquaintance with physical science, and second, several of the fundamental beliefs and mores of the social science community were incompatible with those of science itself. Material was drawn largely from the Vienna Circle, including logical positivists, of course, but also those strongly opposed to positivism, such as Karl Popper. Empirical work carried out in the late 19th and early 20th centuries, along with abstractions drawn from the form of classical Newtonian physics became the core of modern social science methods.

The most elementary concept in quantitative methodology became the variable[1]. From the philosophers of science like Hempel and Oppenheim came concepts such as "nomothetic universals" like F=ma, formulae describing the functional relations among variables, which were seen as the ultimate product of science. Thus, early on, the task of social science was established as the *search for relationships among variables*. Among the earliest methods for examining relationships among variables is Durkheim's use of tables from public records, such as rates of suicide by religious

---

[1] *Variables* are typically contrasted with *constants* in physical science, but there are no known constants in social science, so these are generally ignored or glossed over

affiliation. Later more sophisticated methods of partialing in tabular analysis were introduced by Paul Lazarsfeld, who, while not himself a member, often attended meetings of the Vienna Circle.

But textbook writers were also heavily influenced by non-positivists such as Karl Popper, a non-member but also a frequent attendee at the Circle, who held that the truth of propositions could not be induced from observations, but rather hypotheses could only be falsified or fail to be falsified, but never judged true. In this way, false propositions could be "weeded out" leaving only the true.[2]

A fundamental belief and value of the social science culture, however, was and remains the notion that human beings are special, being endowed with a "free will" which renders exact laws of human behavior impossible to achieve. This belief stands in stark contrast to the view expressed by Ernst Schrödinger, one of the most important scientists of the modern era, that *the first requirement for science is the belief that understanding of the phenomena under study is possible* [7]. Another is the idea that social reality is socially constructed, that is, reality is an ongoing construction of human beings' interpretations of the daily events of their lives [8]. Since social reality is therefore not a "real" natural thing, it cannot (or ought not) be studied scientifically like physical reality, which

---

[2] A position which he later recanted.

is objectively "there.' But quantitative social scientists *behaved* as if social phenomena are objectively real, as evidenced by the social science notion of *validity* presented in every social science methodology textbook: a measurement is *valid* when the results of the measurement instrument correspond exactly to the phenomenon being measured.

The belief that objective reality exists and that we can measure it is generally called "positivism", and many quantitative social scientists probably define themselves as positivists. This is a main point of contention among quantitative social scientists and their qualitative colleagues. Both believe that science is the objective measurement of an existing reality that can, indeed, be measured, but qualitative advocates believe that social phenomena are not objective, existing things, but rather socially constructed concepts that have no objective existence outside our subjectively generated reference frame. Both of these views are in marked contrast to modern physical science, where leading scientists believe that their physical concepts, such as space, time, force, mass, distance and the like are socially constructed, yet are precisely measureable and subject to scientific inquiry.

Social scientists' values were much more compatible with the newly emerging study of inferential statistics, which made only probabilistic statements about the relationships among variables. Thus, the texts incorporated the work of R. A. Fisher, and of Jerzy Neyman and Egon Pearson into the Test Against the Null Hypothesis (TANH), which lies at the core of contemporary quantitative social science practice. Egon

Pearson's father, Karl Pearson, developed the Pearson Product Moment Correlation, or Pearson's r, which gives a measure of the relationship between two variables that is independent of the actual scales on which the variables are measured.[3] This, too, was highly compatible with the deeply held belief of the social science culture that, due to its special ontological status, human variables could not be precisely measured.

Although L. L. Thurstone and others had already shown that human attitudes could be precisely measured (we leave aside for the moment the dispute over whether human attitudes are objectively real or merely subjective constructions), the ever practical and administratively skilled psychologist Rensis Likert showed that his five-point categorical scaling procedure correlated highly with the results of Thurstone's much more elaborate procedures. Practical minded social scientists, who didn't generally believe human attitudes and beliefs could be measured precisely anyway, generally adopted Likert's quick and dirty procedures and precise measurement fell into general disuse.

As coincidence would have it, Karl Pearson's mentor (and Charles Darwin's cousin) Francis Galton, has been credited with having invented the questionnaire, which he administered to prominent scientists of the day. This device

---

[3] Actually, Pearson's r is the cosine of the angle between the variables when expressed as vectors.

has become the principle method by which quantitative social scientists measure variables.

As the use of computers in the university became commonplace, all these procedures, and hundreds of supporting algorithms, were encoded into omnibus general purpose social science software, originally SAS and SPSS, and more recently STATA, R and others. Today, the core of modern social science methodology consists in crude, categorical measurement of variables by means of a questionnaire[4], calculation of the Pearson Product Moment Coefficients (or a more modern non-parametric equivalent) and calculating the statistical significance or lack thereof by means of Tests Against the Null Hypothesis using one of the widely available omnibus computer packages. Advanced research means more of the same, with multiple regression, path analysis or other multivariate correlation-based techniques to discover statistically significant relationships among many variables – some of which may be latent variables discovered by the correlation-based technique of factor analysis. Although there are of course other techniques available, these methods generally make up what the quantitative social scientist would consider the *scientific method* [2].

---

[4] Albeit frequently administered interactively online in the 21st century.

Meanwhile, back at the ranch, the gap between what social scientists believed to be the scientific method and what leading physical scientists were actually doing grew into an insuperable chasm. Not only had physical scientists rejected basic Newtonian concepts like laws ("nomothetic universals") in favor of modern field theory, but they had realized, far before Berger and Luckman had written their first words, that *physical reality* was socially constructed. The notion that observations are dependent on the reference frame of the observer lies at the heart of both relativity theory and quantum electrodynamics. This, however, provided no barrier to the intensification of scientific observation and analysis. Answering the question we left aside earlier, clearly physical scientists believe it is possible to measure subjectively constructed concepts, and to do so with considerable precision.

Chapter 2: The Galileo System

The Galileo System is not a theory or methodology that can be added to the quiver of existing social science researchers. It is rather an *alternative* to contemporary social science theory and method. Based entirely on the model of physical sciences, none of the Galileo procedures can be accomplished using traditional business and social science computer software.

The original Galileo model began by representing cognitive and cultural processes as movements in a high-dimensional continuum – a space that is continuous at all points, rather than the common categorical models like five-point scales and Likert-type scales that make up much of traditional business and social science measurement practice. But human beings – and social scientists and MBA's – do break their experiences into categories. Catpac[tm] and Indstar[tm] are the components of the Galileo model that deal with the categorization – clustering – of human experience. While books describing the original Galileo model in great detail exist, this book is an attempt to combine a good deal of what is known about Catpac and Indstar in a single volume, while emphasizing the continuity of these new applications with the original, continuous, Galileo model.

21

Engineers and computer scientists generally pay little or no attention to how human beings break up the continuous flow of experience into categories. They rather take the view that, in any set of data, there exist natural cleavages – a "ground truth" -- that can be detected by clever algorithms. Their concept of cluster analysis is to devise such clever algorithms and test them against each other on known datasets.

Galileo's concept of cluster analysis, however, focuses on the way in which human beings divide up the continuous flow of sensory data into discrete named chunks. Our goal has always been to understand human cognitive and cultural processes, rather than to find ontological cleavages in sets of data. [1] Catpac's performance depends on various interrelated and non-linear parameters. Early in Catpac's development, experiments were run in which dozens of people read texts and clustered the contents subjectively into categories. Catpac's parameters were adjusted until they most closely matched the clusters derived by the human beings using intuitive methods. *Although they have many practical uses, Galileo, Catpac and Indstar are theories first, and only then methods.*

Introduction: What is The Galileo System?

The Galileo System is not simply an additional method

of analysis or another theory to be added to the social scientists existing quiver. Since it is indeed an intelligent system, people often mistake it for another Artificial Intelligence. Since it provides graphic pictures of attitudes and beliefs, it is easy to mistake for Multidimensional Scaling (MDS) but it is vastly different from that simple tool. Because it can analyze and understand text, you may think it's just another text analysis program, but it is far from that. It's not machine learning, a variant of multivariate analysis, or yet another dreaded set of analytics, although it uses elements of all these specialties.

Galileo is not any special technique or method that fits within the array of tools available to the social scientists. It is, rather, *a rational alternative approach to the scientific study of people and societies.* It is based not on the Galilean/Newtonian model of 18th century science, nor the 20th century positivist model of the social sciences. It is, rather, based on the current model of science as represented by Einstein, Feynman and leading contemporary physical science. It is a radical position whose first principle is that human beings, as an ordinary part of the natural world, do not require a special "kind" of science, but must be studied by the regular methods of regular science.

Galileo is *a powerful theory of cognitive and cultural processes and a well-developed set of tools for observing, understanding and engineering people's thoughts, attitudes, beliefs and behaviors.*

Nearly 50 years ago, I was part of the A. O. Haller's

Wisconsin Significant Other Project — the first ever project that attempted to measure the effects of the expectations of children's "significant others" on their attitudes and beliefs. The data were extremely encouraging but contradicted some of the most fundamental theories in psychology and sociology. The Wisconsin Significant Other Project was initiated by the renowned stratification expert A. O. Haller. It was designed to determine how adolescent children decided how much schooling they planned to attain and what kinds of occupation they hoped to pursue. The study differed from the many others that had gone before by the extensive and precise measurements it used.

The prevailing theory was that all humankind had the same needs and desires, which were implanted into the human mind at birth. Abraham Maslow's Hierarchy of Needs, a popular theory still taught in psychology departments even today, argued that those needs could be placed in a rank order, from the most basic to the highest, which was the same for every human being. How much education children were able to attain, and how high in the occupational status system they ended up was assumed to be determined largely by the children's innate ability and the social and economic resources available to them, measured mainly by their fathers' socioeconomic status[5]. And, of course, there was always the fundamental unmeasureability of human nature, and the child's own free will that lent an irreducible level of unpredictability to the equation.

But that's not what the data were showing.

First of all, the adolescents didn't have the same needs and attitudes — there was great variability in what they hoped and expected to achieve, and what they wanted to be. Rural children didn't end up as farmers because they lacked the resources to be doctors or lawyers — many of them *wanted* to be farmers. The idea that all people have the same built-in needs and desires just didn't stand up in the face of experimental data.

Second, these varied attitudes and beliefs about education and occupation didn't seem to be the result of free will at all, but rather could be predicted very accurately by the *average* of the expectations of the children's significant others-. This indicated two completely heretical results: *first, people's attitudes, beliefs and behaviors are not free choices ungoverned by any lawful scientific process, and, perhaps even more surprising, they are not even made in the individuals' minds, but rather in the social networks in which the individuals*

---

[5] This bias was indicated by the very name of the area of study: *status achievement*, a term which Haller changed to *status attainment*.

lived and then communicated to the individuals in the form of expectations.

Last, but definitely not least, cognitive elements like attitudes, beliefs and the like were not at all unmeasurable but were showing considerable precision in this study.

Extraordinary claims require extraordinary evidence, and the Wisconsin results were soon independently replicated at Ohio State, Texas Tech and Montana State Universities. The evidence for three heretical ideas was growing stronger:

1. *Cognitive and cultural structures and processes can be measured with very high levels of precision.*

2. *Cognitive processes like attitude formation and change take place not in individual minds or brains, but rather in social networks.*

3. *Beliefs, attitudes and behaviors are not the result of free will, but rather obey behave predictably like every other known natural phenomenon.*

Meanwhile, back at the ranch, at the University of Illinois, a long period of intense study of the implications of these three findings began. We also undertook a series of dozens of experimental studies, administering thousands of precise questionnaires over the next four years.

This research revealed that attitudes and beliefs could be represented as points in space. In the case of occupational aspirations, for examples, each occupation could be represented as a point in space. Similar occupations were near each other in this space.

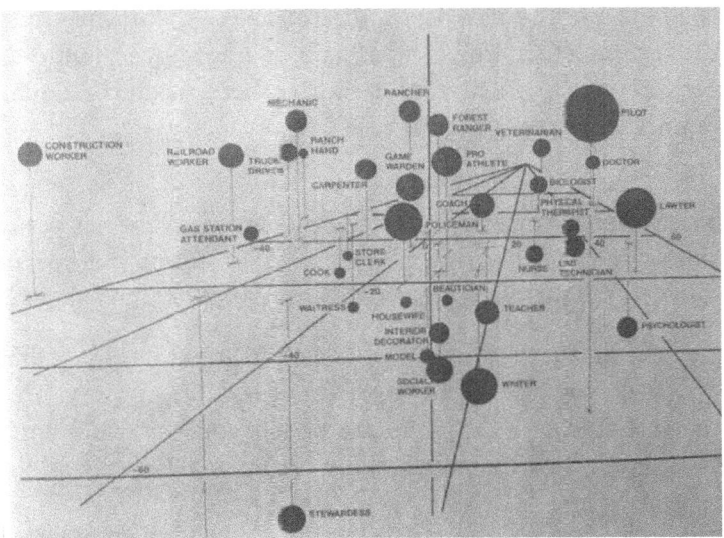

FIGURE 1: OCCUPATIONS IN SPACE

People, too, could be represented as points in the same space, and adolescents would be located closest to the job they expected to attain. Attitude and belief change could be represented as movements in the space. A person who changed his or her job choice could be seen to move through the space from the job they formerly sought to the job they now sought.

At first, we investigated the possibility that this space would be the familiar three-dimensional space of everyday experiences, and that the motions would obey Newton's laws of motion but experiments quickly eliminated this possibility. Using the most precise psychometric measures known – ratio scaled complete paired comparisons averaged over hundreds of respondents – we measured the perceived distances among common objects of all kinds[6].

The results of these experiments forced us to realize that the space is clearly not three dimensional, but has many dimensions — how many, we don't yet know, but recent analysis of Twitter feeds has revealed over 200 dimensions[9].

Secondly, the space is not a simple Euclidean space, but is curved or warped, like the space of relativity theory. This means that changes in attitudes, beliefs and behaviors don't obey simple Newtonian laws, but be modeled by equations

---

[6] I personally spent a great deal of time and energy with pencil, ruler, and compass trying to fit the measured distances among objects into a geometric space with no success until a brilliant undergraduate student named Frank Bellinger showed me that what I was attempting was impossible because the data weren't three dimensional – something I could hardly conceive of at that time. Worse, they weren't even Euclidean.

generalized from Einstein's relativity theory.[7]

The idea that human thought and action could be modeled as a microcosm of the physical universe — and that mathematical equations could describe these processes with great precision — was at first quite remarkable. But understanding that these processes involved highly multidimensional non-Euclidean equations helped explain why human thought and behavior appeared to be so unpredictable and capricious to the unaided mind, even thought they were indeed understandable by advanced science.

Of course, the idea of representing attitudes and beliefs as points in space had been tried before, by Psychometricians using a procedure called *multidimensional scaling*. But these simple models, generally only two or three dimensional, and only precise to within rank orders, were far too simple to account for such complicated processes.

Comparing MDS plots to Galileo space is like comparing a child's drawing of the Big Dipper to a wide field photograph from the Hubble Telescope — they are both pictures of stars

---

[7] We do not expect cognitive and cultural processes to obey Einstein's equations. But The space of relativity is curved and thus non-Euclidean, and experiments show Galileo space is as well, so Galileo equations must use equations describing non-Euclidean spaces

in space, but there the resemblance ends. Because it was so imprecise, the popularity of MDS soon waned, although some analysts still use it on the odd occasion.

Figure 2 shows a highly simplified view of the first three dimensions of the Galileo Space in which the 2016 Presidential Election is taking place (the high-dimensional space is not able to be visualized). It shows five major candidates, several major issues, and yourself, which represent the position of the two respondents (one democrat and one republican) whose attitudes and beliefs the space represents.

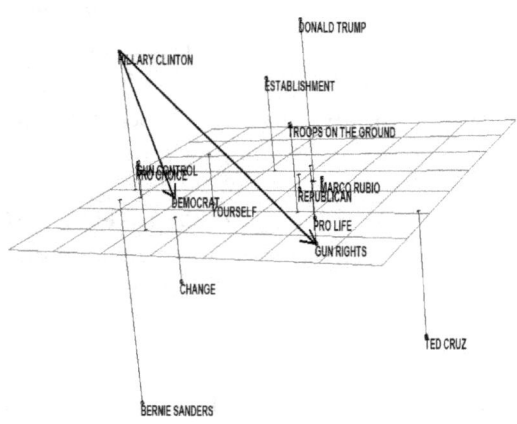

FIGURE 2: CANDIDATES, ISSUES AND SELF-POINT FOR THE 2016 US PRESIDENTIAL ELECTION

Over the past 40 years, we've learned a great deal about this space. We know that the distances among objects in the space are meaningful: objects and people are close to the attributes they exhibit and far from those they don't — Hillary Clinton is close to Democrat, Gun Control, and the Establishment, while Marco Rubio is close to Republican, pro-life and gun rights, for example.

More important, we've learned that we can move these objects through the space in predictable ways. We know, for example, that saying that Hillary Clinton is a Democrat (true) and that she is in favor of Gun Rights (not true) will reposition her close to "*Yourself*," the position of the average of these two voters. In the diagram, the two outer vectors represent this message strategy from *Hillary Clinton* to *Democrat* and *Gun Rights*, and the center vector represents the result of that strategy -- moving from her current position toward *Yourself.* (The simple diagram makes it look as if you might figure out the most effective messages by examining this simple picture, but in reality, the strategies are calculated by equations generalized from relativity theory.)

Equally important is the fact, learned over 40 years of careful research, that behaviors located closer to the self-point (*"Yourself"*) are performed more frequently than those further from the self-point. Figure 3 shows one of the earliest studies showing the relationship between distance of a behavior from the self and the frequency of behavior:

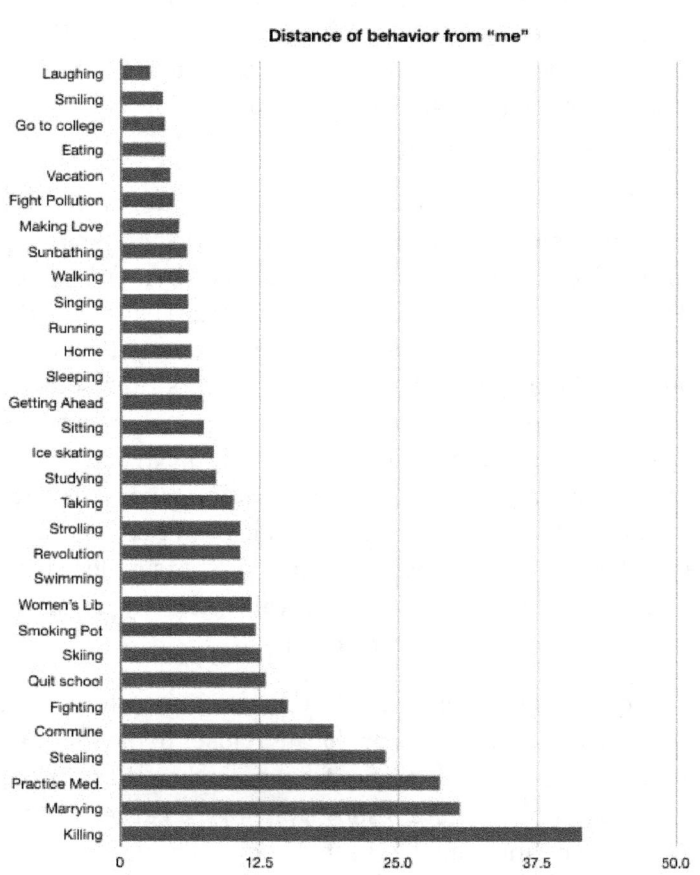

FIGURE 3: DISTANCE BETWEEN SELF-POINT AND BEHAVIORS

This is why we know that moving a candidate closer to the self-point will increase her/his percentage of the voter. *Going to college* is quite close to the self-point, since all these respondents are college students. *Revolution* is also mid-distance, because these data were taken in early April, 1970, when the notion of revolution was driven by events like the Viet Nam War, the Civil Rights Movement, particularly on university and college campuses. *Fighting* is quite far away. The behaviors farthest from the self-point are *fighting, living in a commune, stealing, practicing medicine, marrying*, and, farthest of all, *killing*.

Why is this useful?

When properly implemented, Galileo strategic modeling will always produce the most effective engineering strategy that is possible given the current state of knowledge. No other persuasive technique known can produce a strategy as effective as the Galileo model. No human judgment is required, and the optimal strategy is computed by solving equations in the computer.

The Rand Corporation, after a review of all theories of persuasion in the social sciences commissioned by the US Army, considered the Galileo System the most effective of all:

One of the more interesting approaches to communication and attitude change we found was Joseph Woelfel's ... approach, which is called Galileo. In many ways, Woelfel's theory was the closest that any social science approach came to providing the basis for an end-to-end engineering solution for planning, conducting, and assessing the impact of communications on attitudes and behaviors[10].

Catpac

One of the most important steps in forming a Galileo space is discovering what concepts ought to be included in the neighborhood. An early technique we used was Thurstone's clustering procedure, where concepts possibly relevant to the neighborhood would be written on the back of 4X6 cards, then sorted into eleven piles by tireless graduate students. After each sort, the number of the pile (from 1 to 11) into which each card had been sorted was written on the back of the card. After a hundred or so sorts, these numbers were averaged, and the standard deviations calculated. Those with high standard deviations were rejected because there was little agreement about which pile they belonged in, indicating the concept was probably ambiguous in meaning.

This was a tedious procedure, and Rick Holmes and I wrote a computer program called Johnson (You could call me

Ray, or you could call me Jay, or you could call me R. J., but ya doesn't have to call me *Johnson!*) that did much the same thing relatively automatically, which was a great convenience.

Much of the early testing and development of The Galileo System went on at the Communication Institute of the East West Center in Honolulu. One key study attempted to assess the effects of television on the socialization of children. Samples of children and their parents in five countries were drawn (US, Japan, Korea, Philippines and Great Britain) and the core values of the children and their parents were measured with the Galileo System. Complete video recordings of all the prime-time television for each of the five countries for one full season were obtained, and special recorders were built by Sony that would stop every 15 seconds and allow 14 data coders to record the occurrence or nonoccurrence of each of 150 behaviors — a task that took two full years.

To analyze these data, Rick Holmes and I rewrote Johnson into a computer program called Newton, (Newton, of course, came after Galileo, and the head of the project was Barbara Newton). Newton recorded the co-occurrences of behaviors in each 15-second interval. (Network and cluster analysts will recognize this as the "bag of words" method, which is probably the most widely used quantitative method of text analysis today.) From these co-occurrence matrices, we used multivariate procedures to make "pseudo Galileo" spaces. These spaces turned out not be particularly good compared to "real" Galileo spaces, so we made no use of them.

Meanwhile, back at the State University of New York at Albany, Scott Danielsen and myself, along with a cadre of outstanding students, were developing an artificial intelligence system called Spot. Spot was an artificial neural network, but we didn't know that at the time since that terminology wasn't really standardized until sometime after 1988.

Spot was based on George Herbert Mead's symbolic interaction model, and developed a self-concept based on its role as an assistant to the students in the Department in the use of the Sperry Univac computer. Unlike virtually all other research on artificial intelligence, Spot was not designed as a stand-alone system, but was rather based on the lessons learned in the Significant Other Project: Spot's intelligence developed out of its continuing interaction with its social network, who taught it Exec 8 commands (the operating system of the Sperry Univac computer), and from whom it learned the English language phrases that indicated the user wanted to execute those commands. So, from working with a given student, Spot would learn that "I'd like to edit a file" meant "@ED (filename)" in Exec 8. The intelligence emerged from and resided in the social network that included Spot, not in Spot alone.

In 1989-90, Bill Richards, a pioneer of social network analysis and author of the famed *Negopy* program, spent his sabbatical at the State University of New York at Buffalo, where I had just accepted the chair of the Communication Department, and I took that opportunity to write, with Bill, an alternative model of Spot — this time using the back propagation model which went on to be the foundational algorithm behind what is now called "Deep Learning." We later added feedback loops to Spot to give it some self-awareness in the form of a memory of what it just said and called it Rover. This made it possible to teach Rover to recite "A bicycle Built for Two," which Spot could not. Because Rover was aware of its last utterance, each line of the song could remind him of the next line, so, theoretically, it could sing an infinitely long song. It also made it very difficult to teach, since its behavior was at least partially autonomous, so you couldn't easily predict what it would do or say in every instance.

Although we learned a great deal building Spot and Rover, there were problems with the Deep Learning model that made it inappropriate for a small research organization like ours. First, while it may be appropriate to call back propagation "deep" learning, it might be even more appropriate to call it "slow" learning, since it required extensive training data, and had to pour over those data again and again — typically hundreds or thousands or more times — before it learned anything useful. And it could *over* learn, which made it rigid and unable to deal with anything but exactly what it had learned, and thus far less useful. A great deal of extensive and expensive research over a long time would be necessary to bring it to useful life. Indeed, it took dedicated scientists like Geoffrey Hinton at the University of Toronto, well-funded companies like Deep Mind in London financed by people like Elon Musk and ultimately the financial muscle of Google to make Deep Learning into a formidable tool that could play games competitively.

More important for us, Deep Learning was only in the widest sense anything like actual organic intelligence. First, models had to be designed by humans, who decided how many neurons, how many layers, and the overall general architecture of the network. Secondly, the network did not learn at all like an organic neural network, but rather solved complicated differential equations to estimate the connections strengths (synapses) connecting the neurons to each other. Our interest was never in making a useful applied network, but rather in simulating actual human intelligence as closely as possible. Very few organic intelligences can solve differential equations, and none can solve such an extensive set as quickly as back-propagation demands.

We turned back to our original Spot model, where the architecture of the system was determined on the fly as it learned to carry out the job it was assigned to do, getting better as it worked. This time armed with a much more advanced understanding of neurons, synapses and the mathematics of simulation, we produced better and faster learning systems. We put these newly developed networks into what used to be Newton, now called Catpac, for CATegory PACkage, and, the clustering was so significantly improved that Catpac became a highly useful program on its own merits.

With its new, powerful learning system, Catpac could produce a good approximation of a neighborhood in Galileo space simply from reading text. What's more, its ability to learn in a single pass rather than requiring hundreds or thousands of rereadings of the data made it about six or seven orders of magnitude faster than the Deep Learning model. As computers, the Internet and our technology advanced over the years, we are now at the point where Catpac routinely reads about 30,000 tweets per minute and produces 200 plus dimensional Galileo representations of the main concepts underlying them in seconds[11]. By the time you read this book, it will be much more.

As an example of how it compares to text analysis systems, one recent analysis of the novel Dracula using a highly respected text analysis program from Cambridge University required a full markup of the text — a process that took two full years — and some original programming, followed by weeks of analysis. Of course, concepts not designated in the original text markup could not be detected. A Catpac analysis of the same novel required no markup at all, encompassed every concept and sub concept in the book, and arrayed the main concepts in Galileo space in about 5 seconds — on a MacBook laptop.

Why is this useful?

The past few years have seen the rise of *Big Data, Data Mining* and *Analytics*. In spite of the original hopes, none of these has produced much in the way of notable successes. The notion of big data implied that searching through the immense stores of data collected for random purposes would routinely yield hidden jewels of useful knowledge. The tools for conducting this sifting and winnowing make up Data Mining and Analytics. But in truth, Data Mining tools have been disappointing, requiring considerable specialized expertise, tedious data recoding and analysis. And most require that you know what you are looking for, which is a very hard condition to fulfill. Analytics are with very few exceptions simply public domain statistics routines clothed in public domain plots and charts, presented in at least 4 million colors with new (or not so new) proprietary software.

Catpac, on the contrary, is not just another simple data display utility. Its intelligence is modeled on the Galileo model, where the distances among concepts are modeled as synaptic connections among neurons. It recognizes patterns in text in the same way as human beings detect concepts when reading and can scan through mountains of text on the Internet without recoding, in any language, and detect concepts without pre-specifying what concepts you are seeking. Whatever patterns are there, Catpac will find.

Indstar

From the original Galileo model and the hundreds of thousands of precise measurements over more than 40 years, we learned a great deal about how human thought and action worked — enough to be able to engineer belief, attitude and behavior changes with generalized relativity equations. From work with neural networks like Spot, Rover and Catpac, we learned even more sophisticated ways to construct intelligent systems, which, with Galileo as our guide, we could make ever more human-like. Indstar is a result of that intense research.

Like Catpac, it utilizes a proprietary intelligent engine that emulates human thought, both individual and collective. But, unlike Catpac, it is not restricted to text input but can examine any kind of input data and understand it in a human-like way. Where Catpac can recognize, store and retrieve patterns in textual data, Indstar can detect, store and retrieve patterns in *lists* as would a group of human experts.

One such use is medicine. Indstar can read the records of patients, each of which can be thought of as a list of demographics, symptoms, diagnoses, treatments and outcomes, and detect recurrent patterns in this set of lists. When a new patient presents a certain set of symptoms, Indstar can make diagnoses, suggest treatments along with the likelihood of various outcomes. It can also estimate costs of alternative treatment.

Another obvious and general use is product and service attributes and sales. Indstar can read lists of products people of various demographics along with their features and price points, then make recommendations for any customer depending on their demographics, desired features, price points and the like.

Indstar can also serve as a football coach, reading the history of offensive formations, defensive formations, game statistics and play outcomes to predict what kinds of defenses are most effective against what kinds of offensive formation under various circumstances and vice versus, as well as any other aspect of the games.

Indstar excels at the analysis of social networks, and can learn the communication patterns among any set of people or organizations, so the any one or more persons can be designated, and Indstar will present the others with whom that person or persons communicates about what topics and how often.

Indstar bears a superficial similarity to collaborative filtering, in which a company searches its database for another person most similar to the prospective buyer and suggests products bought by that person. Collaborative filtering is fraught with difficulties, which I won't discuss here except to ask: When was the last time you actually bought a product recommended by Amazon? When was the last time you bought a product recommended by Google search? I rest my case.

Galileo is a comprehensive and powerful intelligent system for observing and influencing human attitudes, beliefs and behaviors. It can serve as a potent instrument of marketing, or an even more effective tool for promoting social and cultural change. It's Catpac instrument is capable of monitoring flows of text across the Internet or any source and providing close to real time analysis of the concepts, beliefs and attitudes underlying those texts. The Galileo model itself allows for precise measurements of attitudes and beliefs, along with a proven mathematical capability for recommending effective engineering interventions. And its Indstar capability is perfect for automated systems that require human-like pattern detection and retrieval capabilities without carbon-based deficiencies like fatigue and bias.

Chapter 3: The Galileo Model of Intelligence

It's not possible to discuss artificial intelligence properly without understanding that there is no consensus on what organic intelligence may be. At the onset of the computer age, the prevailing model of human intelligence was Aristotle's rational actor, a goal-oriented logician who calculated rewards and costs in an effort to maximize his utility (very few scholars of the era would have considered women as a model for intelligence, since the Aristotelian model is severely biased against women.)

Logic was the centerpiece of intelligence among the founders of Artificial Intelligence, and if-then statements stood at the center of the "expert systems" of the time: *if the engine turns, check the spark; if the engine does not turn, check the battery.* Simon, Newell and Shaw's pioneering artificial intelligence program was, in fact, called "Logic Theorist." *Logic Theorist* eventually proved 38 of the first 52 theorems in Russell and Alfred North Whitehead's "Principles of Mathematics," a feat that remains to this day outside the capabilities of all but a tiny minority of highly trained human beings. The term "artificial intelligence" was indeed well coined, since these systems were, in some sense, "intelligent," but without question, *artificial,* bearing little or no resemblance to actual human thought.

This tradition of artificial intelligence continues in the field of computer science, where some of the most popular and interesting examples of artificial intelligence and machine learning produce their intelligence in quite non-human ways. *Deep Learning* for example, involves some processes that resemble organic intelligence, but, at the core, requires the solutions of complex simultaneous differential equations – something which perhaps .000007% of humans now living— or fewer -- might be able to do, given plenty of time and a computer to help them, but something no unaided human being could ever accomplish.

The fundamental goal of Galileo research, from the beginning, has always been to understand the cognitive and cultural processes of actual human beings, and the model of intelligence underlying the Galileo system is taken from the study of human beings.

Organic intelligence in living creatures is implemented by special cells called *neurons*, which generally lie at rest, but can be activated when excited by some external or internal stimulus. When the rods and cones in the retina of the eye, for example, are excited by photons bouncing off a cat, they form a simulacrum of the cat on the retina by the pattern of activated rods and cones struck by the photons. This image is, in turn, communicated to neurons in the brain by *axons* connected to other neurons by *synapses*.

When these patterns recur frequently enough, synapses grow to connect together the neurons that represent the image of the cat, and the organism "remembers" the cat. The memory is contained in the synaptic connections. If one hears the word "cat" while seeing the cat frequently enough, the neurons that represent the image of the cat become connected to the neurons that represent the sound of the word "cat." Thereafter, seeing the cat will excite the sound of the word "cat", and hearing the word "cat" will excite the image of the cat, as will the written or printed word. Still other sensations – purring, meowing, fur, whiskers[8] – will also be represented by neurons linked to each other. Eventually, this extensive network of interconnected neurons will represent a person's concept of "cat" but ultimately the actual concept of "cat" will be stored distributively in the set of all human brains.

Computer Simulated Neural Networks

In an organic intelligence, all this communication among neurons through synapses and axons occurs nearly simultaneously. Even though the electro-chemical impulses that carry signals among the neurons move relatively slowly (about 100 meters per second) they travel simultaneously, and so all the neurons can communicate with all others to which they are connected in a jiffy. Since there are about 86

---

[8] Philosophers and social scientists tend to call these "attributes" while computer scientists prefer "features."

billion neurons in a human brain, with each connected to perhaps a thousand others, the eighty-six thousand billion neural connections yield a staggering capacity to store information.

Galileo

As we've seen, stimuli, sensations, experiences and other objects that are related to each other in human experience are represented in brains as networks of interconnected neurons. The more closely related these experiences are, the more tightly connected they are, and the "closer" they are to each other. The principle Galileo model depends on this proximity model of intelligence to represent the experiences, beliefs, and attitudes of human beings.

As we have also seen, there is no need – and, of course, no possibility -- that all of human experience can be encoded in a single brain, so this knowledge is distributed across all human brains. Within any brain, the neurons are connected directly by tissue – by actual synapses joining neurons together. But neurons in different brains can be connected by communication in social networks, so that concepts stored separately in two spouses' brains, for example, are nevertheless "closer" to each other than concepts in two brains not connected by any communication channels, and activate each other when the spouses communicate.

Potential Typical Applications of the Galileo model

In this section, we describe some potential applications of the Galileo model that are based on proven practices derived from experience with Galileo in a commercial context. Since we have considerable experience with the auto industry, and have access to potential clients in the industry, some of the examples are drawn from the auto industry but are completely general an apply across all markets and industries.

Market Space

In its present simulated form, Galileo software can scan the Internet or other information sources and find patterns which can be arrayed in a high-dimensional space. Figure 4 shows a broad range of autos, ranging from the small, inexpensive Honda Fit to the $200,000 Lamborghini Huracan. Similar cars (e.g., Honda Accord, Toyota Camry) are close to each other in this space. Sports cars in the same price range (Mercedes S class, Porsche Boxster, BMW Z4) cluster at the top right. Specialty cars (Lamborghini Huracan) are at the outer fringes of the space. The concept "yourself" (top center gives the center of the cluster of potential customers: cars closest to this spot sell the highest market shares[9].

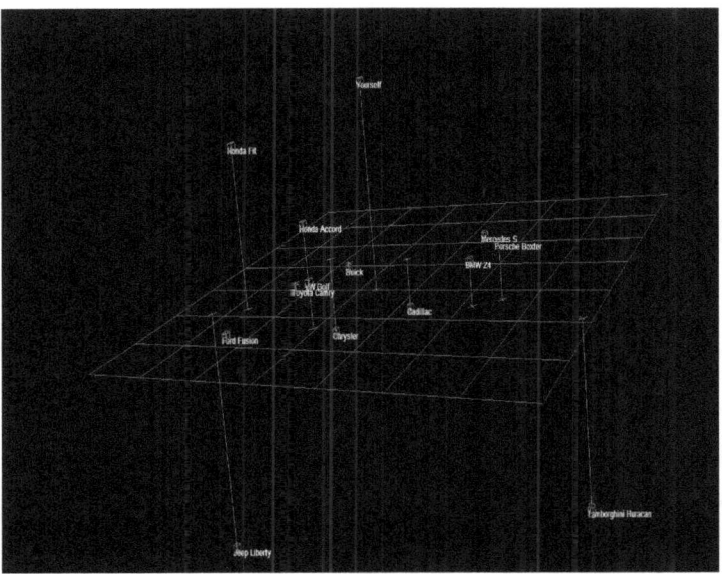

FIGURE 4: AUTOMOBILE MARKET SPACE

Galileo can easily provide a close up of any region in the space. Notice at the bottom left of the picture is the Jeep Liberty, which is a representative of the SUV market. Figure 5 provides a close up of that region

49

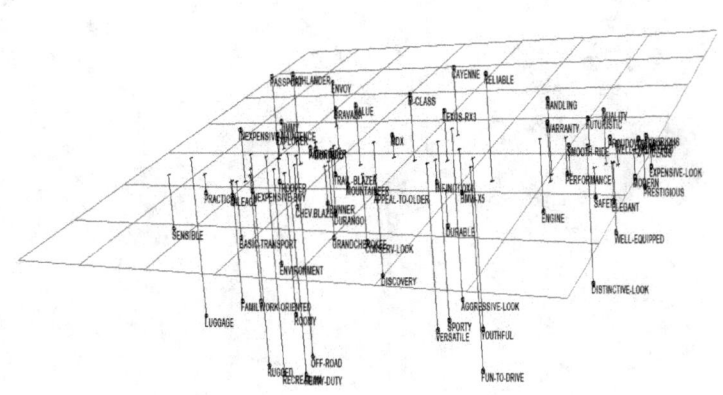

FIGURE 5: CLOSE UP OF THE SUV REGION OF AUTO MARKET SPACE

In the Galileo space, products (cars, in this example) are located near the attributes they embody. Unlike typical measurements, where products either have an attribute or not (smooth, good tasting) or are rated on five-point scales, in Galileo space a product exhibits an attribute to a greater or lesser extent as it is closer or further from that attribute in the space. The attribute may be considered a *field of meaning*.

People may also be represented in the space. They are located closer to the products they buy and use than to those they don't. Market share is greater for products closer to the location of the people in the market (the self-point). In one nationwide study performed for a major manufacturer, the correlation between the distance of several major auto brands (Ford, GM, Chrysler, Toyota, VW) and the self-point and market share of the same brands is -.993[12].

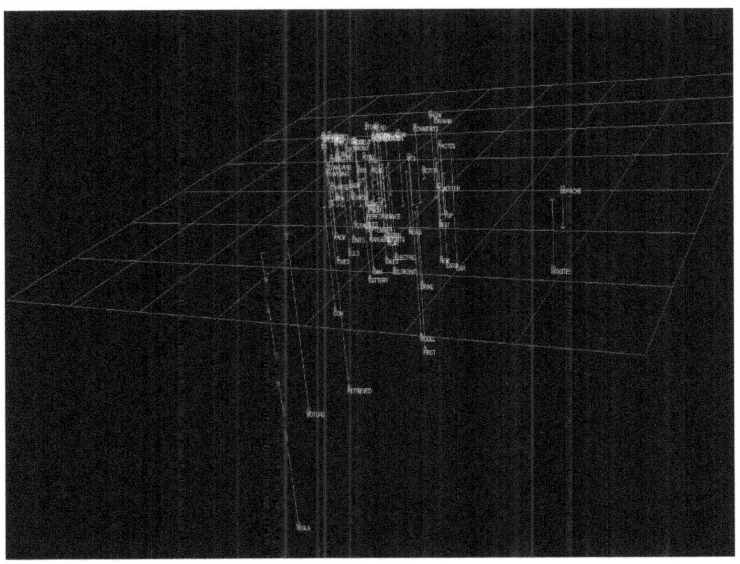

FIGURE 6: TESLA MODE S AND PORSCHE BOXSTER IN GALILEO SPACE

Figure 6[10] shows the neighborhood of Galileo space containing the Tesla Model S and the Porsche Boxster. These are two of the most highly regarded cars in the marketplace, but the Boxster outsells the Model S by almost 10 to 1 (neither car is a high-volume vehicle, which shows once again that how much a car is liked is not a strong predictor of market share.) In the Galileo space, the Tesla Model S is 5.55 units from the self-point, while the Porsche Boxster is only .94 units from the self, which is consistent with the general rule that products closer to the self-point sell higher market share.

Strategic Engineering

Galileo provides not only predictive capabilities, but also provides a powerful and proven engineering capability for moving objects (in this case, car models) through the space to desired locations using mathematical procedures provided by the Galileo model.

---

[10] The map in Figure 6 was made by passing all the text concerning the Tesla Model S and the Porsche Boxster derived from the first two pages of results from a Google search of each name through Catpac and its ancillary programs Space and Strategy. The Catpac model used for this example used only 75 neurons

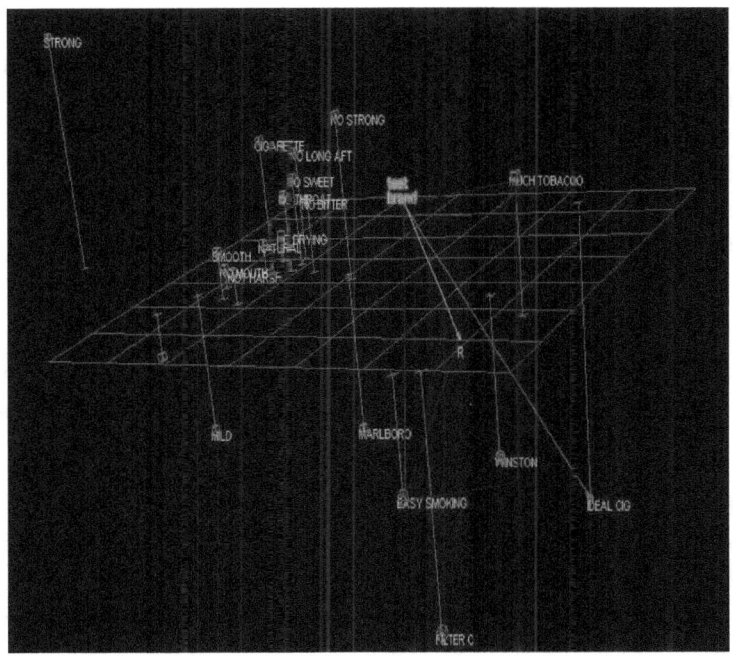

FIGURE 7: OPTIMAL STRATEGY FOR MOVING TEST BRAND TOWARD THE IDEAL CIGARETTE

Galileo's proprietary algorithm calculates the most likely effects of combining different words or concepts in the Galileo space to describe the concept to be moved. Figure 7 shows graphically that the combination of *rich tobacco* and *easy smoking* will move the test brand to the point labeled R in Figure 7; this is the closest to the ideal cigarette that any combination of words can achieve.

In the case of the Tesla Model S, Galileo searched through 373,536 possible combinations of terms and discovered that the combination of *electric* and *Boxster* could reduce the distance between the Model S and the self-point to .11 units from its present 5.55. Even better is the combination *electric Cayman* (the Cayman is the Boxster's hard top twin) which can reduce the distance to .02 units.

Brand Image

The number of potential applications of the Galileo model is too extensive to discuss in a single document, but we should mention brand valuation and politics. In a double blind over time study, Winston smokers and Marlboro smokers were asked to smoke two cartons of Winstons and Marlboros. In one condition, the cartons were regular cartons you might find in a store. In a second condition, the cartons were unbranded plain white boxes containing only the surgeon general's warning.

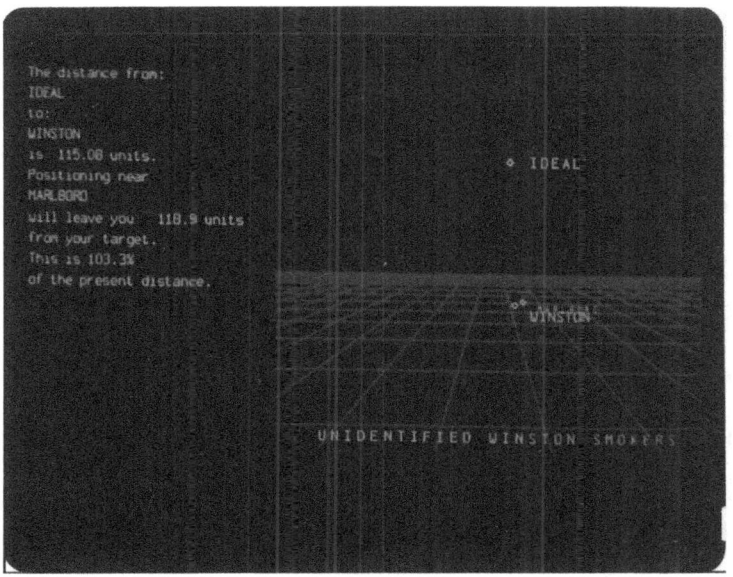

FIGURE 8: WINSTON SMOKERS' PERCEPTIONS OF WINSTON AND MARLBORO IN DOUBLE BLIND CONDITION

Figure 3 shows a Galileo map of Winston smokers' perceptions of Marlboro and Winston when smoked in a double-blind condition. Notice that the brands are perceived by regular Winston smokers to be virtually identical, and almost exactly the same distance from the ideal cigarette. In an unbranded condition, Winston smokers cannot distinguish between the brands.

55

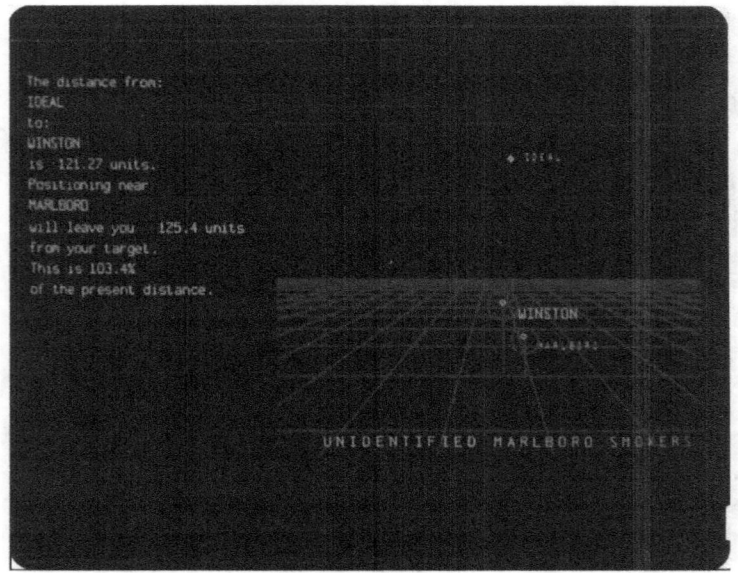

FIGURE 9: MARLBORO SMOKERS' PERCEPTION OF WINSTON AND
MARLBORO IN DOUBLE BLIND CONDITION

Figure 9 shows that, like Winston smokers, Marlboro smokers perceive Winston's and Marlboro's to be virtually identical, and, in fact, see Winston as very slightly closer to the ideal cigarette.

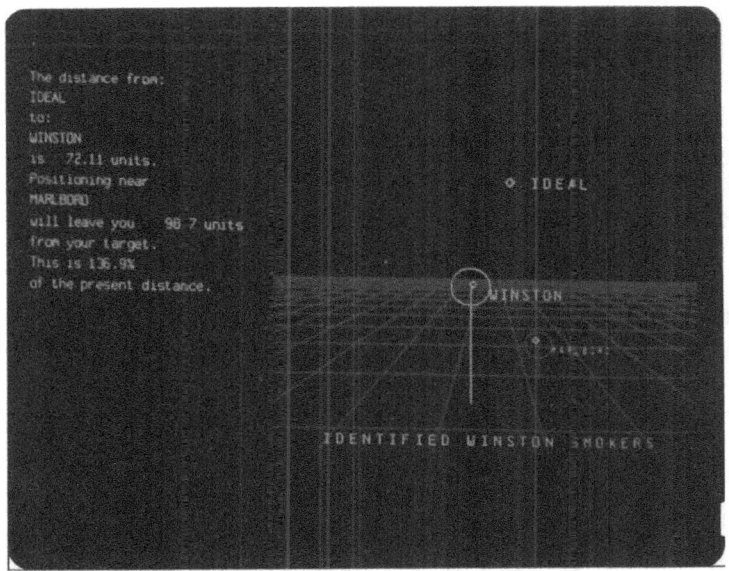

FIGURE 10: WINSTON SMOKERS' PERCEPTION OF WINSTON AND
MARLBORO IN BRANDED CONDITION

Figure 10 shows how Winston smokers perceive the
two brands when they are smoked in their regular, branded
packaging. Notice they see them as quite different, and prefer
Winstons by a great deal, with Marlboros 37% further away
from the ideal cigarette than Winstons.

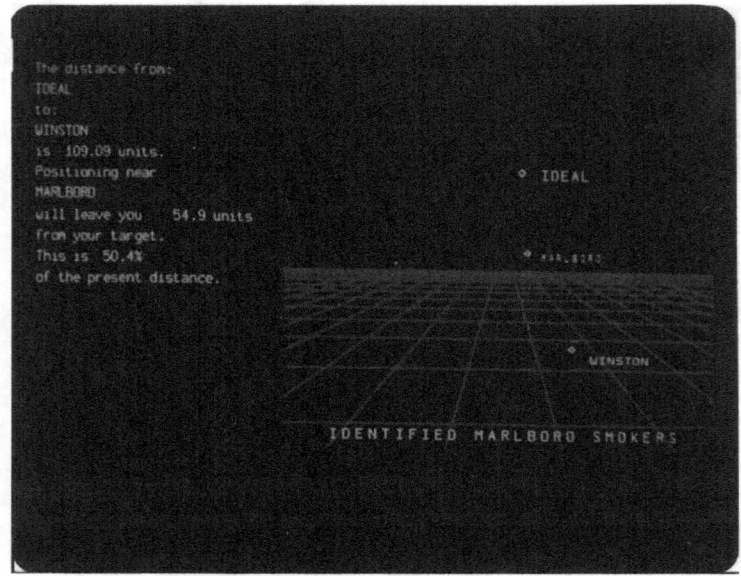

FIGURE 11: MARLBORO SMOKERS' PERCEPTIONS OF WINSTON AND MARLBORO IN BRANDED CONDITION

Figure 11 shows how Marlboro smokers perceive the two brands in their regular packaging. Notice that the effect of the branding is even greater for the regular Marlboro smokers, who prefer Marlboros by fully 50%. Clearly the brand image overwhelms the physiological experience, even though the participants each smoked two cartons – 400 cigarettes. At an average of about 7 minutes per cigarette, this is about 1400 minutes, or just under 24 hours of smoking.

Experiments and Copy Tests

Galileo is ideal for experiments and copy tests. In conventional experiments, it is necessary for the experimenter to understand rather completely what the likely effects of the experiment might be before the experiment is conducted, since measurement instruments need to be devised to detect these results. With Galileo multidimensional representations, this is not necessary since changes in any possible direction can be detected by the Galileo instrument.

Experimental subjects listened to music played through a simple computer speaker system of ordinary quality. While listening they read either of two different messages about the sound they were experiencing. Half the groups chosen at random read this paragraph:

> *Engineers are enthusiastic about the new digital audio format, WH3, which they hope will soon replace the well-known MP3 format. WH3 is capable of much higher levels of compression and so takes up much less space on smart phones and tablets. It also downloads faster and thus is cheaper to use. Although the actual quality of the sound is less like live music, engineers believe the format still has good dynamic range, frequency response and impact.*

The other randomly chosen half read this paragraph:

Music lovers are not enthusiastic about the new digital audio format, WH3, which they fear will soon replace the well-known MP3 format. WH3 is capable of much higher levels of compression and so takes up much less space on smart phones and tablets. It also downloads faster and thus is cheaper to use. However, the actual quality of the sound is less like live music, and music lovers believe users will notice the loss of dynamic range, frequency response and impact.

The exact effects of these different messages are easily captured by the multidimensional Galileo model. The Galileo space for those who heard the engineer's message is shown in Figure 12 while the space for those who heard the music lovers' space is found in Figure 13.

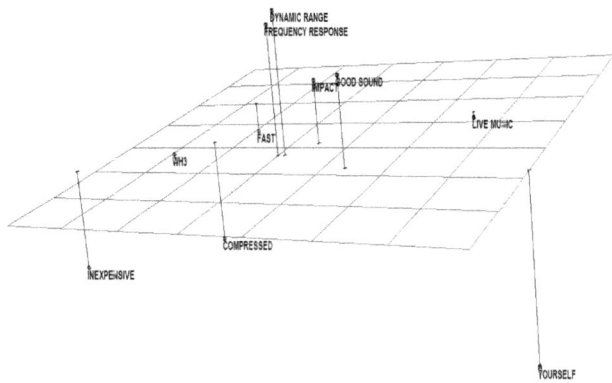

FIGURE 12: GROUP WHO READ ENGINEERS MESSAGE

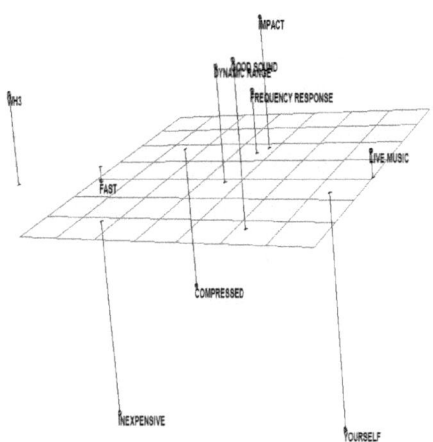

FIGURE 13: GROUP WHO READ MUSIC LOVERS' MESSAGE

Although a complete analysis of the effects of these different messages is beyond the scope of this chapter, it is possible to measure the holistic changes of every concept relative to all other concepts by comparing their locations in the Galileo spaces across the two different groups. Significantly, the distance between the WH3 system and the self-point of those who heard the engineers' enthusiastic message was 257 units, while the distance from the self-point of those who heard the music lovers' disappointed message was 296 units. Clearly the second group liked the WH3 sound considerably less than the first group, although, of course, both groups heard identical sounds.

What this means is that the Galileo system can measure the holistic effects of complex experiences such as advertising campaigns, political speeches, movies, news events and the like with considerable precision.

Politics

The ability of Galileo to measure the holistic effects of complex messages and events, along with its ability to measure the propensity of people to act and to develop effective strategies to influence those propensities makes it ideal for the analysis of political and other information campaigns.

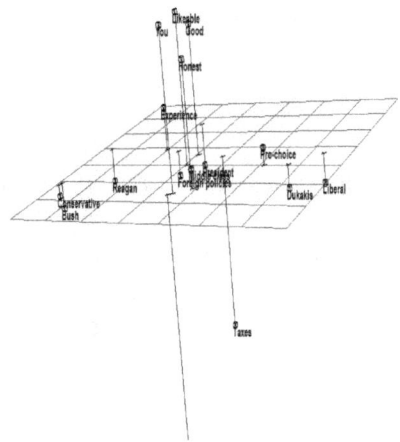

FIGURE 14: GALILEO SPACE OF 2008 PRESIDENTIAL ELECTION

Figure 14 shows the Galileo space of the 2008 US Presidential Election. Notice that the self-point is slightly to left of center of the space. Obama and Biden are closer to the self-point than McCain and Palin, and won the popular vote among this population segment.

It is possible to represent any demographic or other characteristics, of course, such as political party, race, and ethnicity in the Galileo space simultaneously. Flows of information across the Internet can be represented by real time updating of the Galileo space. Optimal strategies for moving candidates and issues closer to the self-point of segments of the population can be generated and update in real time.

Issues and candidates can be tracked through the Galileo space over time. Figure 15

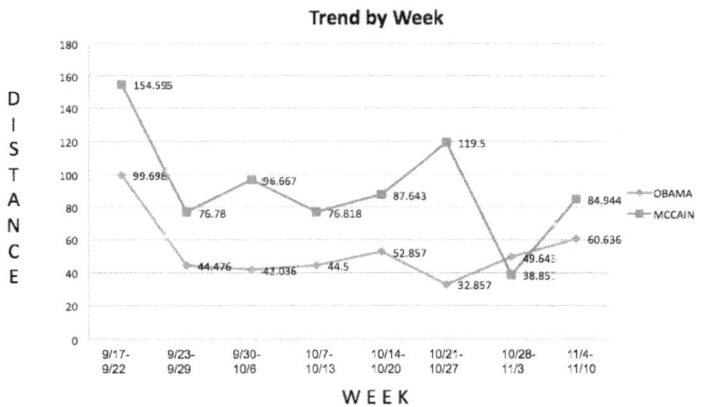

FIGURE 15: DISTANCE OF OBAMA AND MCCAIN FROM SELF-POINT BY WEEK

shows the distance between Obama and McCain from the self-point over the last eight weeks of the campaign.

Advertising and Public Relations Campaigns

Of course, an election is in principle no different from any other campaign to market a product or service. The Galileo model adapts easily to the marketing of any product or service.

Messages

In the Galileo model, anything that changes the pattern of interconnections among neurons is called a *message*. Of course, any change in the values of the synaptic connections among neurons in the neural network implies a corresponding change in the locations of the objects in Galileo space. Messages can be *simple*, referring to only a single pair of concepts, for example, "pigs are beneficial," or *compound*, referencing more than two concepts, such as "pigs are beneficial and attractive."

Galileo theory can predict the effect of these messages on beliefs and attitudes before the fact and allow for precise measurement of the effects after the fact. This gives planners the ability to play "what if" for various prospective scenarios and makes it possible to calculate the most effective messages for creating desired change in beliefs, attitudes and behaviors.

As we've said, the most fundamental rule of neural processes is the Hebb rule: neurons that fire together wire together. That means when attention is focused on any two concepts, the synaptic connection between them is strengthened. This means also that the corresponding "distance" between them in Galileo space is shortened. Similarly, if any three concepts are simultaneously excited, the connections among all three will be strengthened, thus shortening the Galileo distances among them. Within the limitations of attention, we can generalize this to n concepts to suggest the general rule: *when n concepts are simultaneously activated, they will "move" toward their common center in Galileo space.*

Messages and Source Credibility

Each year, Gallup polls nearly a thousand Americans to rate the trustworthiness of common occupations. With the exception of 2001 (the year of the 9/11 attacks) when firefighters topped the list, each year the most trustworthy occupation has been *nurses*. Sadly, for a long time, the least trustworthy occupation on the list has been *members of congress*.

Figure 16 shows the results of an experiment [13] in which a sample of undergraduate students at a large public research university read a message which said "This questionnaire will ask your opinions about the Health Care Reform Act (HCRA), which a committee of nurses said was beneficial and attractive."

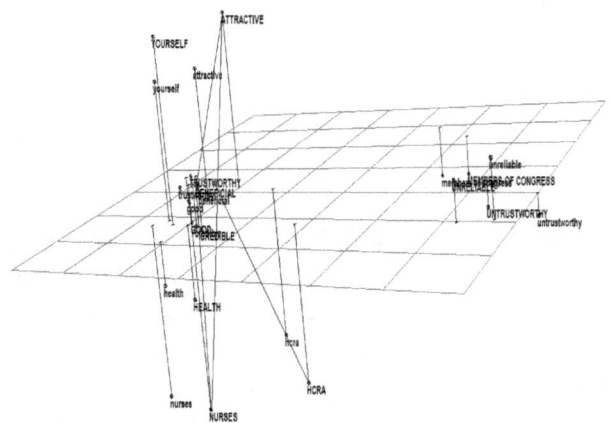

FIGURE 16: EFFECTS OF NURSES' MESSAGE ON HCRA

In this condition, the concepts nurses, HCRA, beneficial and attractive are simultaneously excited. Words in all caps represent the locations of concepts in a control condition which heard no message; lower case represents those who heard the *nurses* message.

The triangle left of center in Figure 16 connects the locations of *nurses*, *beneficial* and *attractive* in the control condition. Following from the general rule above, we predict that the HCRA will move toward the center of the triangle. The line starting at HCRA in the control group and passing through *hcra* in the treatment group (bottom of figure 16) points almost exactly toward the predicted center point.

In the same study, another group of students read a message that said "This questionnaire will ask your opinion about the Health Care Reform Act (HCRA), which a committee of members of congress said was beneficial and attractive." Figure 17 shows the result of that message:

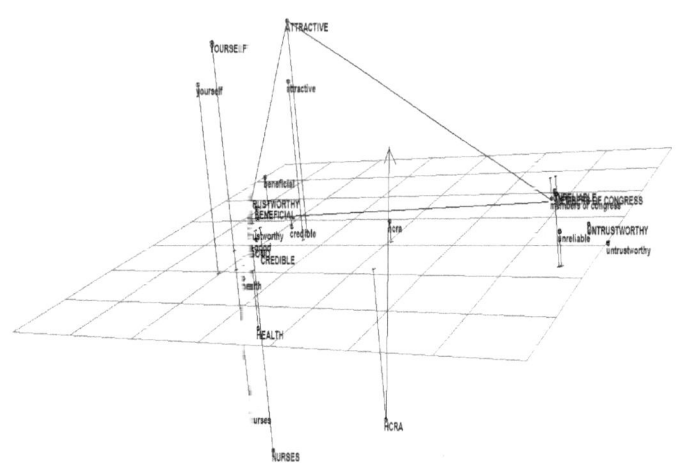

FIGURE 17: EFFECTS OF MEMBERS OF CONGRESS' MESSAGE ON HCRA

The triangle at the top center of the picture connects the locations of the message concepts *members of congress, beneficial and attractive* in the control condition. Again, following from the general rule above, we predict that the HCRA will move toward the center of the triangle. line starting at HCRA in the control group and passing through *hcra* in the treatment group again points almost exactly toward the predicted center point.

Source Credibility

A very important implication of the fundamental rule in Galileo theory is that prior understandings of *source credibility* need to be reconsidered. A very large literature has consistently shown that messages sent by less credible sources have less effect than identical messages sent by more credible sources. Without exception, however these studies have used unidimensional scaling techniques to measure the effects of the persuasive messages. The Galileo system's fully multidimensional measurement system, however, shows not that less credible sources have less effect, but rather that they have *different* effects, in that the actual effects are in different directions. Effects in different directions cannot be simultaneously measured by a single unidimensional scale. The source should not only be considered more or less credible and therefore more or less effective, but rather the source has to be considered an equal part of the message itself. *Two messages delivered by two different sources are different messages, even if otherwise identical.*

Mass or Miracle?

A glance at Figures 16 and 17 will show that the general rule is sufficient to predict the direction in which any given concept is expected to move in response to a given message, but it is not sufficient to predict *how far* in that direction a concept will move. If we continue to believe that the human mind is immaterial, then there is no reason why attitudes cannot change completely and immediately – one might say, miraculously, which is another way of saying cognitive processes are inexplicable by science.

If we remember that concepts are, indeed, clusters of neurons made up of atoms and molecules like all other matter in the universe, we might expect that cognitive processes will have to obey the laws of physics. Those attitudes and beliefs which are based on a great deal of communication history can be expected to be large clusters of neurons, while those formed on the basis of limited communication ought to be smaller clusters. Changing the connection patterns of large neural clusters clearly will require more energy than change patterns of smaller clusters, so some beliefs and attitudes should be expected to be more massive and consequently harder to accelerate than smaller beliefs and attitudes.

Experiment tends to support these ideas. In the HCRA study at the University at Buffalo described earlier, 150 undergraduate students heard a message claiming a fictitious congressional act, the Health Care Reform Act (HCRA) was beneficial and attractive, attributed to either a committee of nurses, or a committee of members of congress, or, in a control condition, no message at all.

By the general Hebb rule, we should expect all five of the concepts in the message – *HCRA, nurses, members of congress, attractive and beneficial* – to approach their common center, and, as we have seen, they do. But the HCRA, which is a fictitious concept[11], should have relatively little mass, and should be expected to move more easily than the other concepts. In the *nurses* condition, the HCRA moves 41.27 units, while nurses moves only 17.69 units, which implies that nurses are almost three times as massive as the HCRA. In the *congress* condition, the HCRA moves 36.37 units, while members of congress move only 2.66 units, implying that members of congress are over ten times as massive as the HCRA. Since congress is much more widely covered in daily media, it is no surprise that *congress* is 6.65 times as massive as *nurses*, and fully 13.67 times as massive as the fictitious HCRA.

Pigs in Space

In an early study at an Eastern Polytechnic Institute, Barnett [14] estimated the inertial masses of four synonyms, pig, hog, boar and swine. Problems with Barnett's pioneering experiment allowed him to estimate only the masses of pig, hog and swine, but he found their masses correlated 1.0 with the Thorndike-Loge index of frequency of occurrence of the terms in English. A replication of Barnett's study 38 years later

---

[11] Many subjects would be expected to conflate the HCRA with the many proposed bills in congress to revise or repeal the Affordable Care Act, so we would not assume there is no information history, only that it is likely much less than the other concepts.

at the University at Buffalo, State University of New York was able to estimate the masses of all four synonyms. The correlation of the estimated masses with the original Thorndike-Loge index of frequency of occurrence was .995, while the correlation with another contemporary index of frequency of occurrence was .983. This is substantial evidence that the resistance to change of concepts is governed by the amount of information out of which their synaptic connections have been established, and not to any subjective feelings or emotions the individual might attach to them.

The implications for applied persuasion are clear: not all beliefs and attitudes are similarly easy or difficult to change, and the degree of resistance to change – their inertia mass – is determined by the amount of information out of which they have been formed. Whether and how much any given belief or attitude will change when exposed to a given message is not under the voluntary control of the recipient.

These findings are consistent with earlier research on significant others (cf. Chapter 2) that showed attitude stability depended not on affective feelings about the aspirations or the source, but only on the amount of information out of which the attitudes were formed.

Chapter 4

Galileo for Cross Cultural Comparative Research

Neuroscientists generally understand that concepts have a physical basis as clusters of interconnected neurons, although sociologists and anthropologists understand that cultural concepts transcend individual brains, and are clusters of neurons distributed across many brains interconnected by social networks (Woelfel 2010). In the Galileo model, these physical structures are modeled as points in a multidimensional Riemann space, and the interconnections among them are represented by distances or separations.

Identifying which concepts are relevant to any given topic begins with a few interviews on the topic of interest. Once the concepts have been selected, distances among them (representing the interconnections) are measured by a survey instrument that pairs 10-20 key terms, each with every other, followed by computer analysis using the Galileo software.

The method is particularly useful in comparing cognitive features from one group to another (men to women), and/or over time. The visual outputs (maps) are convenient for rendering the computer's complex mathematical calculations clear to those interested in the results. Figure 18 shows how the attitudes and beliefs of men and women about basic emotions can be represented in a single picture. Circles represent men's opinions, while triangles represent women's. Concepts that are close together are tightly connected -- notice at the far right of the picture, Happiness and Joy are close together. Concepts that are far apart are less connected; notice that Happiness (far right top) is very far from Hate (far left bottom.)

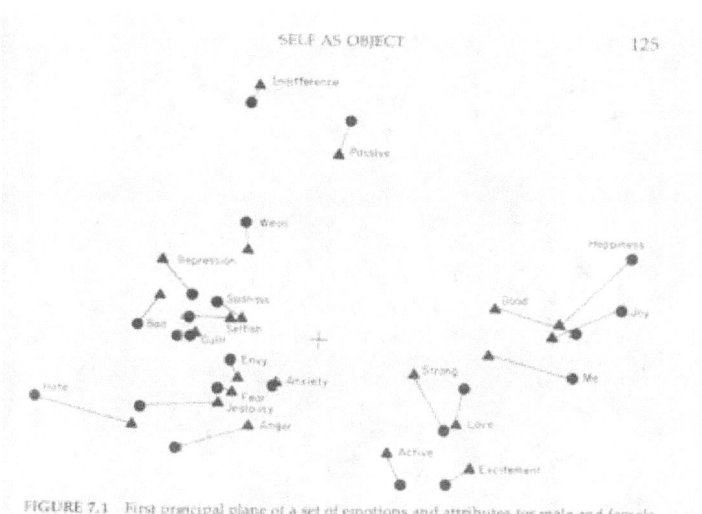

FIGURE 7.1 First principal plane of a set of emotions and attributes for male and female university students plotted on joint coordinates. Triangles represent female respondents; circles represent male respondents.

FIGURE 18: MALE (CIRCLE) AND FEMALE (TRIANGLE) EMOTIONS

Figure 19 shows that the Galileo model can be used cross culturally without difficulty. It shows the relationships among several basic human emotions by native speakers of

75

English, Korean and Hindi projected onto the same space:

Notice at the bottom center the words for "yourself" in the three languages. The distances from these "self-points" to the emotions indicate the degree to which those emotions are connected to the respondents; e.g., people who are closer to Happiness are happier.

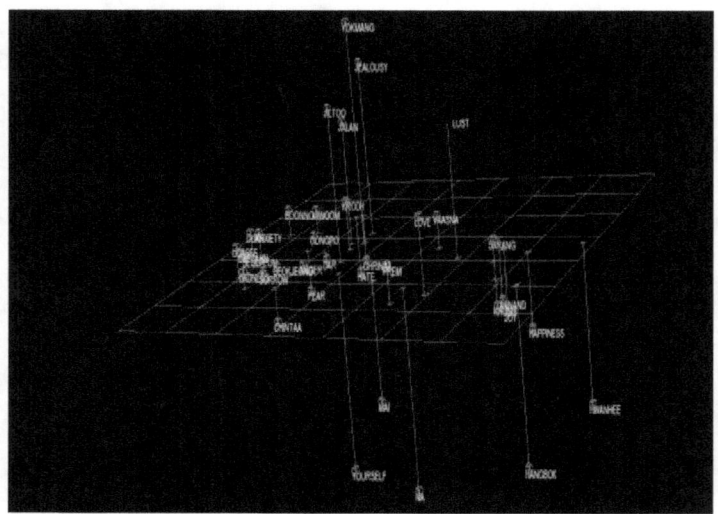

FIGURE 19:EMOTION NAMES IN ENGLISH, HINDI AND KOREAN

Because the first three dimensions of the Galileo Riemann space can be plotted, the procedure bears a superficial resemblance to Multidimensional Scaling, which caused considerable confusion early on. But the Galileo coordinates go far beyond a quick and simple picture. Table 7.1[12], for example, shows the numerical values of the distances between men and women and the basic emotions:

TABLE 7.1

Emotional Attitudes by Sex

		Love	Hate	Anger	Joy	Envy	Fear	Jealousy	Happiness	Sadness	Anxiety	Excitement	Indifference	Depression	Selfish	Guilt	Strong	Bad	Active	Weak	Good	Passive	N
Male	$\bar{X}$	25	239	146	28	121	135	142	28	97	81	52	161	108	118	131	38	128	33	131	50	133	33
	σ	6	53	55	4	29	27	34	5	15	12	20	43	28	22	27	6	29	4	28	13	36	
Female	$\bar{X}$	24	160	85	32	85	103	84	41	90	78	39	142	99	109	105	45	105	41	97	45	93	46
	σ	6	25	9	8	12	14	12	7	10	10	5	26	14	13	13	7	13	7	11	9	11	
Combined	25	193	110	30	100	116	109	36	93	80	45	150	103	113	116	43	114	37	111	48	110	79	
	4	27	24	6	14	14	16	5	8	8	9	24	14	12	14	5	14	4	13	7	16		

By reading across the rows, it's possible to derive a quantitative profile of the emotional condition of men and women (Woelfel and Fink 1980). Similarly, the resulting data (means matrices, coordinates, etc.) are flexible and fruitful for subsequent analyses---whether disaggregated by gender of respondent or by terms for men and women included in the survey instrument. Besides the mapping function, these results can also be used to fashion persuasive messages more effectively, based on local concepts and worldviews. It is also important to keep in mind when using the maps (rather than the numerical results), that they can only reflect three of the cognitive dimensions measured.

---

[12] Reprinted from Woelfel & Fink, 1980, p.126.

A Proven Technology

Galileo is a proven technology. Galileo has been used successfully to map election campaigns[15, 16] convergence of the cultural beliefs of immigrants[17], media[18, 19] perceptions of TV content[18] occupations[20-25], translation accuracy[26, 27], facial expressions[28] , tourism[29, 30], cultural development[31], ideologies[32], social networks[33-35], school violence[36], forests[37-39], music[40], organizational climate[41], sex roles[42] , people[43], land management[44, 45], technology clusters[46], legal policies[47], wildlife management[48] ,international images[49], aircraft[50], cognitive states[51], pigs[14], international aid[52], sustainable agricultural patterns[53], organ and tissue donation[54], and smallpox vaccination[55], and a large body of unpublished commercial research[12].

Artificial Intelligence implemented on von Neumann architecture is interesting and helpful in understanding organic intelligence as well as providing useful practical tools. But soon we will be constructing systems with real intelligence that should in no way be considered "artificial". The morality of dealing with actual intelligent systems implemented in silicon has been treated largely in science fiction, but it will be essential that such considerations be at the forefront from the very start of development.

Chapter 5:

Procedures for Precise Text Analysis:

The idea that the complexity of human experience is underlain by a small set of thema or archetypes stretches to antiquity, as is clearly visible in Plato's Ideas and Aristotle's Categories[56]. Early social scientists like Charles Spearman and L. L. Thurstone hoped to be able to reduce test scores of individuals to a small set of underlying "factors" defining human intelligence. The growth of "big data" has intensified efforts to find convenient methods of reducing massive amounts of data to smaller, simple, meaningful and useful representations.

One of the most popular methods has been categorization or clustering, in which various elements in the data are considered similar enough to be treated alike. Among the earliest such schemes is VALS, (values, attitudes and lifestyles), launched by SRI in 1978[57]. Subsequently many more VALS-like schemes followed. Although substantial differences exist, most follow a simple two-stage model, where individual's values, attitudes and lifestyles are measured in some way, similarities among them are calculated by some algorithm, and then these measurements are reduced to a smaller number of categories through some clustering or factor-analytic scheme. All the individuals in each resulting category are considered similar enough to be dealt with identically by marketers, advertisers, and other interests.

Although Spearman's and Thurstone's methods were based on individually administered tests, and early VALS-like systems were usually based on questionnaire-type data, by far the largest body of data available today consists of text, and the focus of contemporary "big data" research has shifted to the analysis of text, and this article will focus on text as well. Many programs and algorithms for text analysis exist, but it is convenient to divide them into two broad types: rule-based analysis and propinquity based analysis. Rule based text analysis consists of those type of analysis based on linguistic, syntactic, grammatical or other theoretical schemes of analysis. These methods differ among themselves widely, but share the notion that the correct interpretation of language depends on some rules or schemes, either learned or genetic[58, 59]. We will not be concerned with rule-based clustering methods in this paper.

Propinquity based analysis, on the other hand, considers grammar, syntax, rules and the like to be devices invented by analysts rather than the basis on which individuals generate and interpret language. Instead, it assumes that words tend to become associated in meaning simply because they frequently occur "close" to each other in discourse. In this view, when President Nixon says, "...the American People deserve to know whether their president is a crook. Well, I am not a crook..." the "not" is meaningless, and the concepts of "Nixon" and "crook" are forever associated thereafter.

Propinquity analysis consists of two steps: first, the measurement of the propinquities or distances among the elements, and second, the procedures by which the elements are divided into categories or clusters once the propinquities have been established. We will consider here two broad types of clustering algorithms: hierarchical or "hard" clustering methods, which assign each element into its one "best" category, and context-sensitive algorithms, which can assign the same element into one of several possible categories depending on context.

Measuring Propinquity

The oldest form of propinquity analysis is co-occurrence analysis. The most elementary form of this type of analysis is what Danowski[60] calls the "bag of words" approach[13]. In this form of analysis, the number of times words co-occur in the same "bag" – e.g., document, page,

---

[13] Woelfel refers to the bags as "cases" or "episodes", and the Catpac software and manual refer to the "bag of words" method as "case delimited mode."61.   J. Woelfel, *Artificial neural networks for cluster analysis.* (RAH Press, Amherst, NY, 2009), 62.   J. Woelfel, Journal of Communication **43** (1), 63-80 (1993).

episode, utterance, etc. – is counted, and a matrix of frequencies of co-occurrence is computed. This co-occurrence matrix is the basis of all further analysis. Early examples of such programs are Danowski's Wordij,[63] Newton, which constructed co-occurrence matrices based on the co-occurrence of 150 behaviors in 15 second intervals of prime time TV shows in five countries;[18] and Catpac,[62] which has been used very extensively in a large number of substantive areas worldwide.

Although each of the various types of text analysis has its supporters, direct comparisons of different analysis routines on the same data are rare. In this paper, two of the most widely used similarity models – a co-occurrence model and a neural network model – are directly compared and contrasted on the same text. Further, certain important difficulties associated with the mathematical analysis of both methods are discussed, and an alternative model is presented.

Some examples:

Consider the following simple text:

Blue

I only had three dreams in my life I remember, and those none too clearly. I dreamed I went hunting with my father, and it was the only dream I ever dreamed in color. I don't remember anything about it, except the colors of the leaves and trees where we were hunting, and, if you made me swear to it, I couldn't in all honesty say I was sure we were hunting, or that my father was with me. I just remember the colors in the woods.

-1

I actually dreamed another dream in color, but only one color – blue. It was in the yard of our house, and it switched to my uncle's cottage in Sunset Bay. There were flying saucers flying around the yard/bay, and they were blue and they were very compelling.

-1

I remember another dream about an attic in a house, which I now think might be my mother's parents' house, and I confuse that with another dream about a huge house with many hidden and mysterious rooms and stairways, but it's terribly vague, and not in color. The emotion is there, though, and it's an emotion of strangeness and discovery, a tingling emotion, as are the other two dreams I remember.

-1

Walter Mosley's Blue Light made me think of these dreams. He's a far better writer than I am, and his book starts with rays of blue lights on a much more cosmic scale than my dreams – rays of blue light from Neptune crashing through the sun to earth –

awakening questions in frogs and murder in young men. I still have no idea what his blue lights might be, but I know that I know the blue lights from my childhood dreams. I know the blue lights, and I can still feel them.

-1

When I was a teenager I made a device that emitted blue light. It was like the cardboard tube from a roll of paper towels, with a blue light bulb, like an old fashioned Christmas tree light – a Mazda – in it. I connected it to a cardboard box with a few resistors and capacitors wired in no particular way. I knew it had no effect on anything, but I called it a "synapsifier", a fictitious device that worked on the synapses of the human brain. I check the frontispiece of Walter Mosley's book, and find the copyright is 1998. My synapsifier filled my attic room with blue light over fifty years earlier – about the time of Blackboard Jungle, plus or minus. So I had the blue light thing early, but I didn't use it.

-1

This text consists of five paragraphs delineated by the marker "-1". Each marker delineates one "bag of words" or "episode"." A simple cluster analysis utilizing the co-occurrence method[14] yields the dendogram shown in Figure 1:

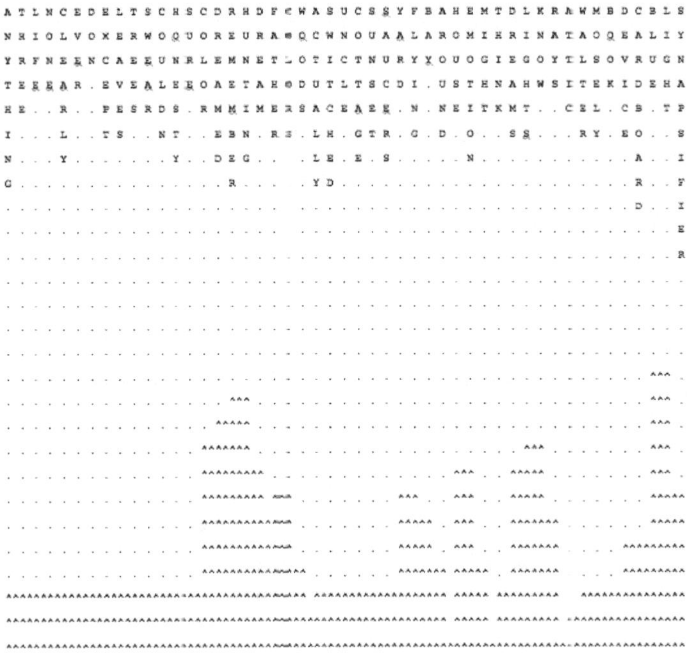

FIGURE 20: SIMPLE CO-OCCURRENCE ANALYSIS

---

[14] All analysis in the following examples were performed using CatpacIII™ from The Galileo Company.

Figure 20 shows a fairly shallow dendogram with only eight levels produced by the Johnson's Hierarchical Clustering algorithm in CatpacIII. Nevertheless, there is enough signal in the result to show clearly the father/hunting episode, the color flying saucer dream (although the saucers are omitted), the Walter Moseley book and the "synapsifier device, both of which are sub-clusters of the Mosley/synapsifier cluster. There is also a fifth cluster, the dreams/rays/lights cluster.

A significant improvement on the "bag of words" approach came from Danowski[64] who first introduced the "sliding window" approach. His approach introduces a better measure of propinquity, in that words that are close together will co-occur more often in the sliding window than words that are further apart. Specifically, in a two-word window, only contiguous words will co-occur. In a three-word window, contiguous words will co-occur twice, while words separated by one word will co-occur once. In a seven-word window, contiguous words will co-occur six times, words separated by one word will co-occur five times, and so on until words separated by five words will co-occur once.

Figure 21 shows the results of a co-occurrence analysis of the same text, this time using a moving window of size three, (the size Danowski recommends), Figure two does not appear to be an improvement on the

FIGURE 21: CO-OCCURRENCE MODEL WITH 3 WORD SLIDING WINDOW

```
H F E T D L N C C I D E R D A E L I S C H S C W A S U C S B F S Y A A H T L K M W M B R B L C D S
U A M H R I O L O B R V E O N X E R W O O U O O C W N O U A L A A R T O H I N I A O D A L I A E Y
N T O R E F N E L B E E M N Y C A B E U N R L O T I C T N Y Y U R O T U I G O G L S O Y U G R V N
T H T E A E E A O F A R E . T E V E A L E E O D U T L T S . I C D U I S N H W H T E K S E H D I A
I E I E M . . R R M M . M . H P E B R D S . R S A C E A E . N E . N C E K T . T E L . . . T B C F
N R O . S . . L . E . . B . I T S . . N T . S . L H . G T . G R . D . . . S . . R Y . . . . O E S
G . N . . . . Y . I . . E . N . . . . . Y . . L E . E . . . S . . . . . . . . . . . . . . A . I
. . . . . . . . . . . . R . G . . . . . . . . Y D . . . . . . . . . . . . . . . . . . . . R . F
. . . . . . . . . . . . . . . . . . . . . . . . . . . . . . . . . . . . . . . . . . . . . D . I
. . . . . . . . . . . . . . . . . . . . . . . . . . . . . . . . . . . . . . . . . . . . . . E
. . . . . . . . . . . . . . . . . . . . . . . . . . . . . . . . . . . . . . . . . . . . . . R
. . . . . . . . . . . . . . . . . . . . . . . . . . . . . . . . . . . . . . . . . . . . . .
. . . . . . . . . . . . . . . . . . . . . . . . . . . . . . . . . . . . . . . . . . . . . .
. . . . . . . . . . . . . . . . . . . . . . . . . . . . . . . . . . . . . . . . . . . . . .
. . . . . . . . . . . . . . . . . . . . . . . . . . . . . . . . . . . . . . . . . . . . . .
. . . . . . . . . . . . . . . . . . . . . . . . . . . . . . . . . . . . . . . . . ^^^ . . .
^^^ . . . . . . . . . . . . . . . . . . . . . . . ^^^ . . . . . . . . . ^^^ . . .
^^^ . . . . . . . . ^^-^^ . . . . . . . . . . ^^^ . . . . . . . . . ^^^ . . .
^^^ . . ^^^ ^^^ ^^-^^ . ^^^ ^^^ ^^^ ^^^ ^^^ . ^^^ ^^^ ^^^ . ^^^ ^^^ . . . . ^^^ . ^^^ .
^^^ . . ^^^ ^^^ ^^-^^ . ^^^ ^^^ ^^^ ^^^ ^^^ . ^^^ ^^^ ^^^ . ^^^ ^^^ . . ^^^ ^^^ . ^^^ .
^^^ . ^^^^^ ^^^ ^^-^^^^ ^^^ ^^^ ^^^ ^^^ ^^^ ^^^^^ ^^^ ^^^ ^^^^^ . ^^^ ^^^^^ ^^^ . ^^^ . .
^^^ . ^^^^^ ^^^ ^^-^^^^ ^^^ ^^^ ^^^ ^^^ ^^^ ^^^^^ ^^^ ^^^ ^^^^^ ^^^^^ ^^^ ^^^ ^^^^^ ^^^
^^^^^^^^^^^^^^^^^^^^^^^^^^^^^^^^^^^^^^^^^^^^^^^^^^^^^^^^^^^^^^^^^^^^^^^^^^^^^^^^^^^^^^^^^^^^
```

simple "bag of words" co-occurrence model for this particular text, yielding more and smaller two-word clusters such as /dreams, remember/,/emotion, dream, / hunting, father/, /Walter, Mosely/blue, light/cardboard, might/ and the like. the number of levels is also reduced to six.

Increasing the window size to seven, does increase the depth and detail substantially, yielding more than twice as many (15) levels, four and five-word clusters, and even sub-clusters within clusters.

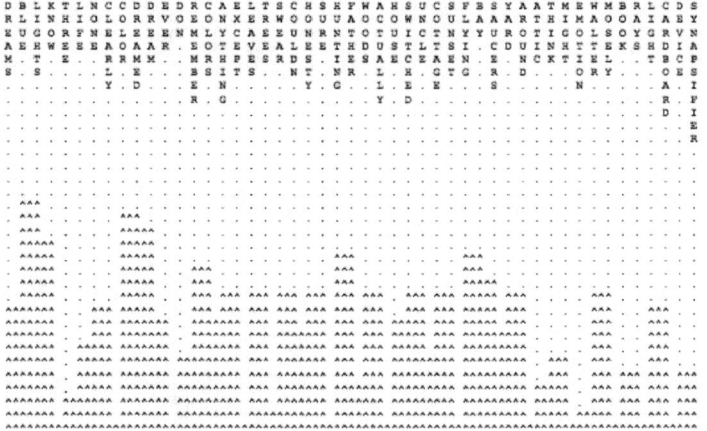

FIGURE 22: CO-OCCURRENCES WITH 7 WORD SLIDING WINDOW

More recently Woelfel[62] introduced an artificial neural network into text analysis. This network was implemented in Catpac™, which was originally a program which computed co-occurrences using the "bag of numbers" approach, where the beginning and end of each "bag" was determined by codes embedded in the text. Later, Catpac implemented Danowski's sliding window method, and, in the late 1980's, introduced a single pass unsupervised neural network to measure the propinquity relationships among the words in the text.

Initially, the network creates a set of artificial neurons, one for each word in the text. Then it "activates" those neurons whose associated word is in the sliding window at each iteration. The "connections" (degree of closeness or propinquity) of those neurons that are co-present in the window are then incremented. As the network grows, however, and connections are established among the neurons, activation of neurons in the window can result in the activation of other neurons not in the window, which are positively connected to those in the window. Connections among all these neurons are also incremented[15].

What this means in practice is that the propinquity of nodes is not established simply on the basis of pairwise co-occurrences, but on the basis of both direct pairwise relations and complete n-way indirect relationships among all the nodes. In practice, the result is deeper, more finely detailed relationships among the nodes, as shown by deeper, more detailed dendograms and perceptual maps.

---

[15] Complete operation of the network, including forgetting, normalization and other issues is not discussed here. For more detail, see Woelfel, 1993.)

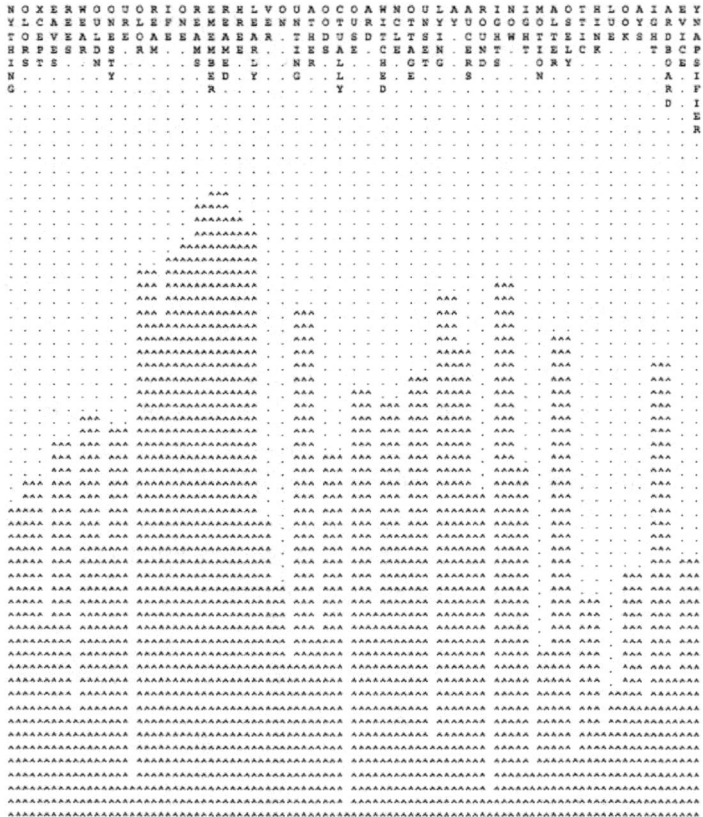

FIGURE 23: NEURAL NETWORK WITH 7 WORD WINDOW

Figure 23 shows that the neural network with a 7-word window clearly finds the deepest, most finely articulated structure of any of the methods tried so far. With 33 distinct levels of clustering and many deep clusters subdivided into smaller sub-clusters. What may be lost in the mix, however, are the four original episodes – the three dreams and the Moseley book and the synapsifier.

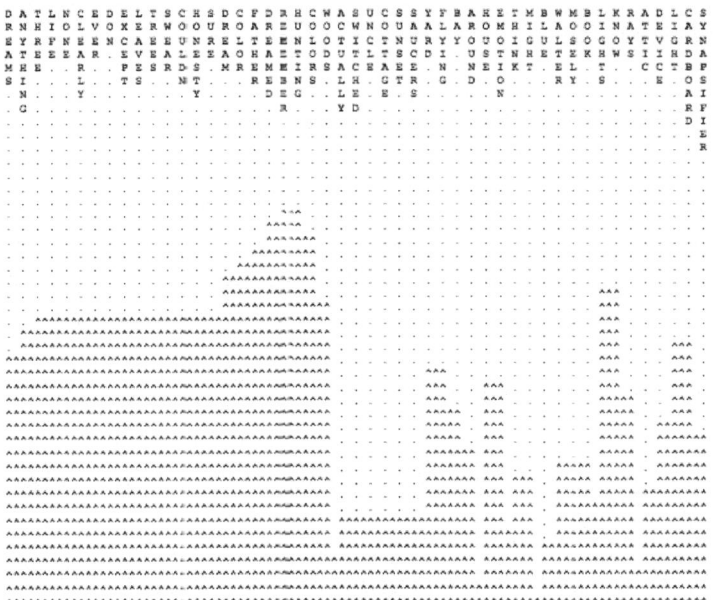

FIGURE 24: NEURAL NETWORK WITH "BAG OF WORDS" OR EPISODES

Figure 24 shows the neural network solution using the original episodes or bags. The three dreams and the Moseley/synapsifier episode are clearly visible in the solution, although the detailed picture of sub-clusters is lost. Generally, if the episodes or "bags" are not meaningful, the moving window neural network is the method of choice. But if the episodes or bags are substantively meaningful, the episodic of "bag of words" approach should be preferred. Since the actual run-time for the software is usually only a second or two, there is, of course, no reason both procedures shouldn't be tried and reported.

One of the instances in which the "bag of words" or episodic method appears more appropriate is shown in the case of medical diagnosis and treatment. Each disease may well be considered a category or cluster that contains symptoms, diagnostic tools, diagnoses, treatments and outcomes. Figure 25 shows the symptoms and suggested diagnostic tools associated with several diseases in the Merck Manual Online[65].

ABDOMINAL PAIN
WAVES OF DULL PAIN WITH VOMITING
INTESTINAL OBSTRUCTION
ULTRASOUND
-1
ABDOMINAL PAIN
COLICKY PAIN THAT BECOMES STEADY
APPENDICITIS
STRANGULATING INTESTINAL OBSTRUCTION
MESENTERIC ISCHEMIA
ULTRASOUND
-1

ABDOMINAL PAIN
RECURRENT

FIGURE 25: SYMPTOMS, DIAGNOSES, TREATMENTS AND OUTCOMES

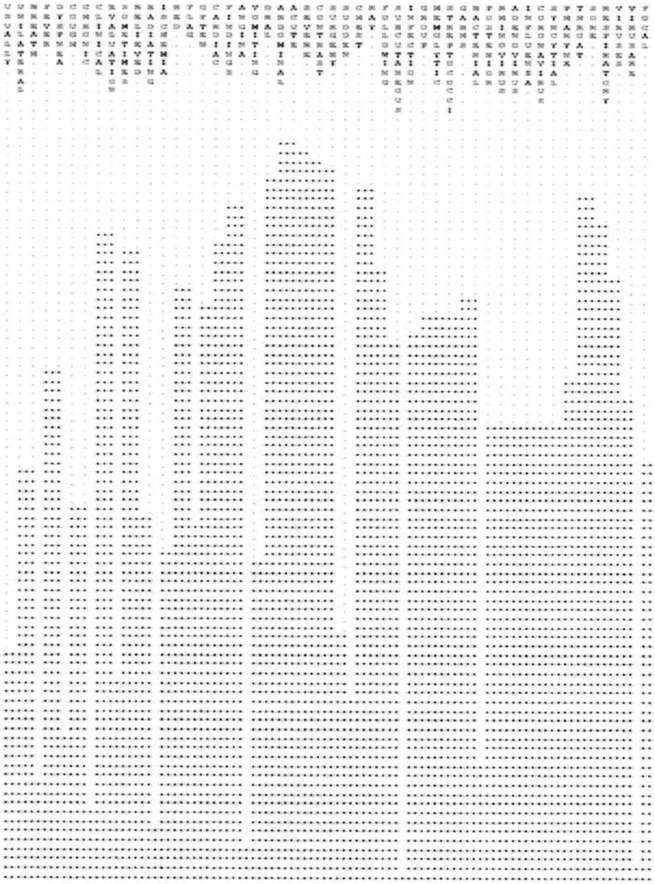

FIGURE 26: NEURAL NETWORK WITH EPISODIC OR "BAG OF WORDS"
ANALYSIS (MERCK DATA - ABRIDGED)

Since the Merck data consists of a list of symptoms, diagnoses and proposed diagnostic procedures, it would seem that the episodic or "bag of words" approach would be appropriate, since a moving widow would necessarily lump symptoms, diagnoses and treatment strategies of one disease with its neighbor in the data file every time the moving window slid across the end of one disease and the beginning of another.

But Figures 25 and 26 reveals a deeper problem: the hierarchical clustering methods discussed so far all share the same shortcoming: they require that every element be placed into one and only one "best" cluster. *Vomiting*, for example, is clustered with angina in Figure 26, but clearly vomiting is a symptom of many different diseases. Similarly, as Figure 25 shows, *abdominal pain* occurs in many diseases, and obviously can't be classified into only one best cluster.

Since the same symptoms can and do occur in several diseases, any clustering method that requires that each element be assigned to one and only one "best" cluster is inappropriate. Fortunately, the neural algorithm can discover content-sensitive clusters that allow elements to belong to more than one cluster at the same time, and to different clusters depending on context.

Because the neural network constructs a network of neurons through not only direct links, but through all n-way indirect links as well, it can calculate the degree to which any given neuron will be activated when any other subset of

95

neurons is activated. This model is implemented in Indstar™, a neural network for non-hierarchical cluster analysis[66]. Basically, Indstar can answer the question "what other elements will reside in a category that includes any given set of elements?" So we can query Indstar by activating the neuron corresponding to *abdominal pain*, and it will respond by activating all the other elements that are (sufficiently) linked to it:

Enter as many items as you want, Doctor. -1 when done		
ABDOMINAL PAIN		
	And the winners are...	Activation
	ABDOMINAL PAIN	1.0000
	ACUTE	0.0052
	SEVERE	0.0052
	ELDERLY	0.0026
	CT WITH ORAL CONTRAST	0.0102
	SURGERY	0.0052
	INFANT	0.0026
	MILD	0.0026
	INCONSEQUENTIAL	0.0026
	HIV	0.0026
	IMMUNOSUPPRESSANTS	0.0026
	CORTICOSTEROIDS	0.0026

FURTHER TESTING	0.0026
VISCERAL PAIN	0.0051
VAGUE	0.0051
DULL	0.0051
NAUSEATING	0.0051
POORLY LOCALIZED	0.0051
UPPER ABDOMEN	0.0026
STOMACH	0.0026
DUODENUM	0.0026
LIVER	0.0026
PANCREAS	0.0026
FOREGUT	0.0026
LOWER ABDOMEN	0.0025
DISTAL COLON	0.0025
GU TRACT	0.0025
HINDGUT	0.0025
ACUTE WAVES OF SHARP CONSTRICT	0.0026
RENAL COLIC	0.0026
BILIARY COLIC	0.0026
ULTRASOUND	0.0076
WAVES OF DULL PAIN WITH VOMITI	0.0025
INTESTINAL OBSTRUCTION	0.0025
COLICKY PAIN THAT BECOMES STEA	0.0025
APPENDICITIS	0.0025

STRANGULATING INTESTINAL OBSTR	0.0025	
MESENTERIC ISCHEMIA	0.0025	
RECURRENT	0.0025	
ULCER DISEASE	0.0025	
GALLSTONE COLIC	0.0025	
DIVERTICULITIS	0.0025	
MITTELSCHMERZ	0.0025	
SHARP, CONSTANT PAIN, WORSENED	0.0025	
PERITONITIS	0.0025	
TEARING PAIN	0.0025	
DISSECTING ANEURYSM	0.0025	
SUDDEN ONSET	0.0025	
PERFORATED ULCER	0.0025	
RENAL STONE	0.0025	
RUPTURED ECTOPIC PREGNANCY	0.0025	
TORSION OF OVARY	0.0025	
TORSION OF TESTIS	0.0025	
SOME RUPTURED ANEURYSMS	0.0025	
FLAT AND UPRIGHT ABDOMINAL X-R	0.0025	
UPRIGHT CHEST X-RAYS	0.0025	

FIGURE 27: INDSTARTM ANALYSIS OF NODES CONNECTED TO ABDOMINAL PAIN.

Clearly, as Figure 27 shows, *abdominal pain* is a symptom of alternative possible diseases depending on, among other things, whether it is acute, severe, visceral, vague, dull, nauseating, and/or poorly localized, and classifying it as a member of its one "best" disease is a serious error. Which disease that symptom is indicating in any given case is determined by the context, which will consist of other symptoms and test results presented. Figure 28 shows how Indstar determines to which category *abdominal pain* should be assigned in the context of an additional symptom, *colicky pain that becomes steady*:

Enter as many items as you want, Doctor. -1 when done	
ABDOMINAL PAIN	
COLICKY PAIN THAT BECOMES STEADY	
And the winners are...	Activation[16]
ABDOMINAL PAIN	1.0000
ACUTE	0.0052
SEVERE	0.0052
CT WITH ORAL CONTRAST	0.0102
SURGERY	0.0052
VISCERAL PAIN	0.0051
VAGUE	0.0051

---

[16] For reasons of space, elements with activation values below .005 are omitted.

DULL	0.0051
NAUSEATING	0.0051
POORLY LOCALIZED	0.0051
ULTRASOUND	0.0101
COLICKY PAIN THAT BECOMES STEA	1.0000
APPENDICITIS	0.0050
STRANGULATING INTESTINAL OBSTR	0.0050
MESENTERIC ISCHEMIA	0.0050

FIGURE 28: INDSTARTM NEURAL ACTIVATIONS FOR ABDOMINAL PAIN AND COLICKY PAIN THAT BECOMES STEADY

Figure 28 shows that including the additional symptom *colicky pain that becomes steady* into the context of *abdominal pain* changes the categories to which it "belongs". Several possible diagnoses have been activated by the inclusion, such as appendicitis, strangulated intestinal obstruction, mesenteric ischemia and the like. Several suggested diagnostic tools have also been activated, such as ultrasound and CT with oral contrast. Further, many possible diagnoses and symptoms have been *deactivated* by the additional context. Which diagnostic tools to use and which disease ultimately turns out to be correct will depend on still more contextual cues. What is important, however, is that *the category to which any given element "belongs" is dependent on context and cannot be settled once and for all by a simple hierarchical clustering scheme.*

Conclusions

Hopes to reduce the complexity of human experience to a small set of meaningful categories date to antiquity. The earliest thinkers attempted to discover these underlying archetypes by reasoning and developed many theoretical schemes – a practice which continues unabated today. Beginning in the early 20th Century, mathematical and later computer-based techniques devolved. Most of these involve the measurement and calculation of similarities among elements by some scheme, followed by a mathematical reduction of the rank of the similarities to some underlying set of factors, clusters or other archetypes.

A wide variety of such computational techniques for cluster analysis continue to exist side by side, but this paper has shown that the method by which the similarities among elements is measured and/or calculated is of fundamental importance, and so not all existing techniques ought to be considered equivalent. Methods appropriate for some types of data may be unsuitable for others and vice versus.

Of even greater importance, the idea that a clustering scheme can be devised in which each element can be assigned to its one and only "best" category is shown to be frequently impossible due to the effects of context. A context-sensitive neural network model, Indstar, is presented which makes it possible to classify elements based not on some "inherent"

101

meaning, but on the basis of the context in which they are experienced.

Chapter 6

Precise Procedures for Non-Hierarchical Cluster Analysis[17]

Although scholars in different fields disagree about the origins of Cluster Analysis, clustering procedures can be traced back at least to Charles Spearman's 1904 work on the factors of human intelligence[67]. Today Cluster Analysis is a mainstay of multivariate research in fields as diverse as sociology, communication, psychology, management, social network analysis, computer science, biology, astrophysics and others. Since 2000, use of cluster analysis has increased greatly, and, since 2004, the preponderance of use has shifted from mathematics and psychology to management and engineering.[68]

The overwhelming majority of cluster analyses in virtually all fields is hierarchical (sometimes called "hard" clustering), which we use here to mean a nested set of clusters

---

[17] Star Trek is a trademark of CBS Studios Inc. Catpac, Galileo, ThoughtView and Indstar are trademarks of The Galileo Company. SPSS is a trademark of SPSS, Inc.

in which each item is assigned to one and only one "best" cluster. Some procedures (sometimes called "soft clustering" or "fuzzy clustering") recognize the possibility that some items might belong to more than one cluster, and calculate coefficients that indicate the degree to which any item belongs to any cluster.[69]

All these procedures treat the meanings of clusters and their members as characteristics independent of context. Human beings, however, classify objects differently based on the context in which they occur. The word "mustang", for example, when presented in a context including Arabian, Palomino and Quarter Horse, will be classified as a horse, but, when presented in a context that includes Camaro, Challenger and Barracuda, will be classified as a car. Although the classical model of human perception and reasoning, based on Aristotle's syllogistic model, is itself categorical and hierarchical, recent research indicates quite strongly that actual human cognitive processes are more accurately described by a non-hierarchical network model[70, 71].

Whatever might work in biology or astrophysics, hierarchical cluster analysis, and more recent "soft" clustering models, are not appropriate for emulating the processes by which human beings cluster their perceptions. Clearly, classifying "mustang" as either a horse or a car, as hierarchical or hard clustering does, is wrong or at least incomplete, and it is equally wrong to say, as does soft or fuzzy clustering, that it is x% horse and y% car.

An Example

A text file consisting of 12 non-exhaustive categories (cars, stinging insects, space shuttles, aircraft carriers, rocket scientists, sports implements, carbonated waters, Star Trek spaceships, ocean predators, works of Igor Stravinsky and horse racing tracks was constructed. Each of these categories contained items that could legitimately be included in at least two or more categories, as Figure 29 shows:

CARS	AIRCRAFT CARRIERS	HORSES
BARRACUDA		CLYDSDALE
	ENTERPRISE	
MUSTANG		ICELANDIC
	HORNET	
CAMARO		MUSTANG
	WASP	
FIREBIRD		PAINT
	INTREPID	
COUGAR		PALOMINO
	SARATOGA	
JAVELIN		PINTO
	YORKTOWN	
AMX		QUARTER HORSE
	MIDWAY	
CHALLENGER		SHETLAND PONY
	CORAL SEA	
		TENNESSEE WALKING HORSE
		THOROUGHBRED

SPACE SHUTTLES	SPACE SHIPS	ROCKET SCIENTISTS
INTREPID	DEFIANT	GODDARD
ATLANTIS	ENTERPRISE	HERMANN OBARTH
COLUMBIA	CHALLENGER	WALTER DORNBERGER
ENDEAVOR	INTREPID	
ENTERPRISE	LEXINGTON	WILLY LEY
CHALLENGER	YORKTOWN	WERNHER VON BRAUN
	GODDARD	
SPARKLING WATERS	**SEA CREATURES**	**STRAVINSKY WORKS**
SARATOGA	BARRACUDA	FIREBIRD
EVIAN	SHARK	THE RITE OF SPRING
PERRIER	KILLER WHALE	PETRUSHKA
PELLIGRINO	SEAL	THE RAKE'S PROGRESS
FIJI	SQUID	
GLENLIVIT	OCTOPUS	THE NIGHTINGALE

		OEDIPUS REX THE FLOOD CONCERTO IN D
STINGING INSECTS	**SPORTS IMPLEMENTS**	**HORSE RACE TRACKS**
HORNET	JAVELIN	SARATOGA
WASP	HAMMER	HIPPODROME
KILLER BEE	SHOT	BRIGHTON
HONEY BEE	DISCUS	GALWAY
YELLOWJACKETS		CHURCHILL DOWNS
MOSQUITO		AQUADUCT
TICK		

FIGURE 23: EXAMPLES OF OVERLAPPING CATEGORIES

Thus, Barracuda can be a car, and/or an aquatic creature; Mustang can be a car and/or a horse; Firebird is both a car and a work by Stravinsky. Of course, the car Mustang and the horse Mustang are not the same thing, but only the same word referring to two different objects – a problem which does not occur in, say, astrophysics or biology, but does occur

107

whenever human judgments are involved, as, for example, in marketing, social network analysis, text analysis, data mining and so on. While it may be argued that such an array of multiple and disparate categories ought to be avoided from the outset, this is exactly the kind of situation that arises when scanning data streams from the Internet.

Conventional hierarchical clustering models have problems with this ambiguity. Figure 30 shows the result of a Catpac[18] analysis of the data in Figure 29. Catpac[62, 72] is one of the most widely used text analysis programs in the social sciences, and although it's distance measures are derived from a sophisticated unsupervised neural network algorithm, its clustering algorithm is a standard diameter method hierarchical clustering method taken from Johnson's original Bell Laboratories algorithm.[73]

Catpac has three major components. The first is a simple parser that breaks text down into its constituent words and discards unwanted words (such as articles, prepositions and the like) and counts the remaining words and breaks the

-----

[18] All analysis reported here were conducted using Catpac III. Although alternate programs, such as Danowski's Wordij or SPSS's Hierarchical Cluster Analysis would yield somewhat different results due to their different methods of calculating inter-item distances, all would encounter the same problem due to the hierarchical clustering algorithms they employ.

text into *episodes*. An episode is either one of a series of sliding windows *w* words wide that slides through the text, creating a new episode each time it slides, or a set of words demarcated in the text by symbols (-1) inserted by the user (Since these data represent distinct categories, the latter procedure was used for the analyses reported here.)

The second component is a proprietary unsupervised neural network algorithm. When a given word occurs in an episode, an artificial neuron representing the word is activated (assigned a value of 1). Once all the words in the episode have been activated, every neuron whether in the window or not is polled to determine its activation value; before the first episode, of course, all activation values will be zero. For all possible pairs of neurons, the synaptic connection between them is adjusted as a function of their activation values and a user-adjustable learning constant This means that neurons that are simultaneously active will tend to become tightly connected. Ignoring considerable computational detail, the result is a square symmetric matrix of synaptic connection weights.

The third component of Catpac is Johnson's hierarchical clustering algorithm. This algorithm searches through the synaptic weight matrix for the largest value. This is the most similar pair of neurons, and the words they represent become the first cluster. The rows and columns representing these two neurons are then replaced by a single row and column representing the midpoints between the original pair. The reduced matrix is then searched for its largest value. If the largest value is a pair representing the now combined vectors of the original pair and another neuron, then the word representing this new neuron is added to the original cluster. If not, a new cluster is begun. This continues until the matrix vanishes.

Figures One and Two show that, on the surface, the hierarchical clustering algorithm has done a reasonably good job of detecting the twelve clusters. The fact that the hierarchical solution is a fair approximation of the data, along with the general belief that human phenomena are too fuzzy and evanescent to be described precisely, probably account for the lasting popularity of hierarchical algorithms over the years. On closer inspection, however, problems emerge. The first large cluster (far left of Figure One) is clearly the car cluster, correctly including *Camaro, Cougar, Mustang, Firebird, Barracuda, Javelin* and *AMX*, but erroneously leaving out *Challenger*. *Challenger* is one of the most ambiguous concepts in the data, being a member of the car cluster, the space shuttles cluster, and the Star Trek space ships cluster. Curiously, *Challenger* ends up in none of these clusters, but instead is weakly attached to the cluster of Stravinsky's works at the right of Figure One.

FIGURE 30: HIERARCHICAL CLUSTER ANALYSIS OF 12 CATEGORIES

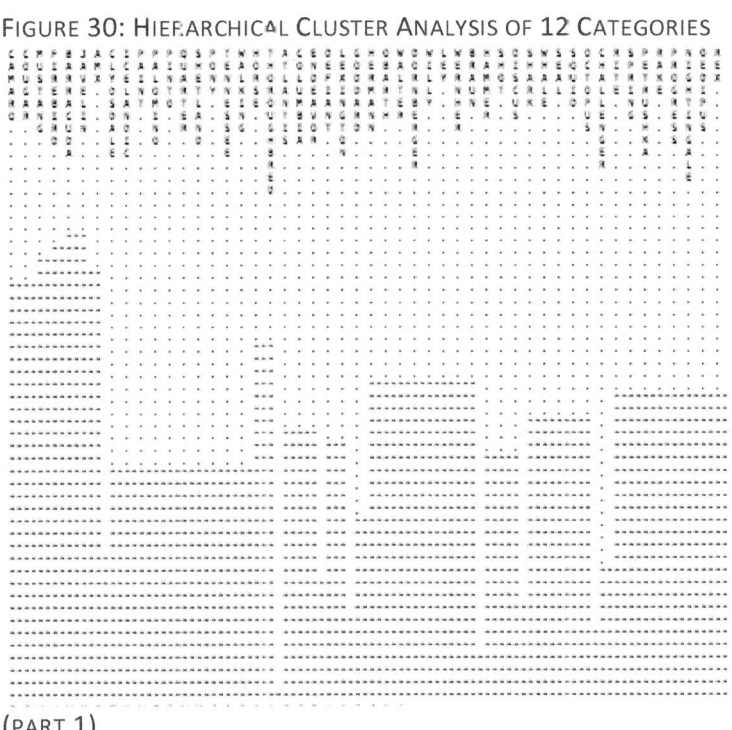

(PART 1)

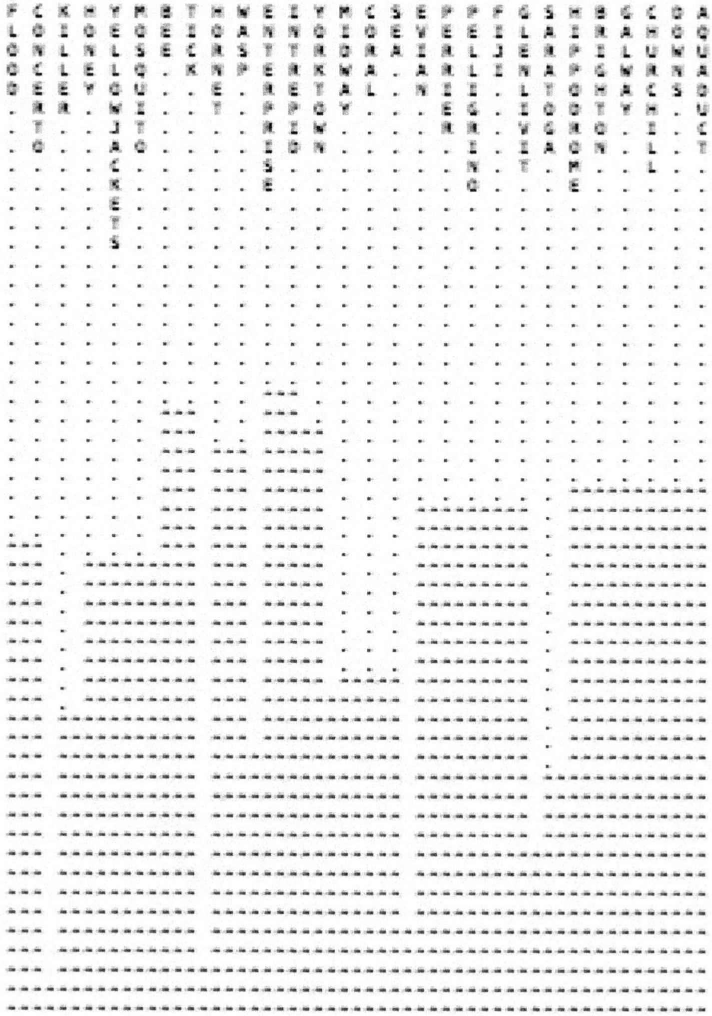

FIGURE 31: HIERARCHICAL CLUSTER ANALYSIS OF 12 CATEGORIES (PART 2)

The second cluster, to the right of the car cluster in Figure 30, is clearly the horse cluster, however it omits *Mustang*, which has already been assigned to its one best cluster, cars. Notice that, about two thirds of the way down the dendogram, the car cluster and the horse cluster merge into a single cluster, which is clearly a mistake based solely on the co-occurrence of *Mustang* in both clusters.

The third cluster, immediately to the right of the horse cluster, is the space shuttle cluster, which correctly includes *Atlantis, Columbia* and *Endeavor*, but omits *Intrepid, Enterprise* and *Challenger.* As we shall see, Intrepid and *Enterprise* have been assigned to the aircraft carrier cluster, and, as we have already seen, *Challenger* has been classed as a work of Stravinsky. The fourth cluster, immediately to the right, includes the two Star Trek space ships *Defiant* and *Lexington*, but erroneously omits *Enterprise, Challenger, Intrepid* and *Yorktown*. These two clusters then merge due to the overlap of *Enterprise* and Challenger in both the space shuttle and Star Trek space ship clusters, but neither of these is, in the end, classified as members of either cluster.

Once any element has been assigned to its one best cluster, it is of course no longer available to be assigned into other clusters where it rightly belongs as well, so *Goddard*, assigned to Star Trek space ships, is missing from the rocket scientists cluster, *Javelin*, assigned to cars, is missing from the sports implements cluster, *Barracuda*, also assigned to cars, is missing from the Sea Creatures cluster, Firebird (assigned to cars) is missing from the works of Stravinsky cluster, Hornet and Wasp are missing from the Stinging Insects cluster (assigned instead to aircraft carriers), Saratoga is missing from both the aircraft carrier cluster and the sparkling water cluster (assigned instead to horse racing tracks).

A Neural Algorithm for Non-Hierarchical Clustering

It is interesting to note that, while the hierarchical clustering algorithm in Catpac is not capable of disambiguating the overlapping terms, the neural algorithm by which Catpac calculates the distances among the items contains all the information that is needed to account for the different meanings of the terms in different contexts. Each pairwise synaptic weight in the resulting matrix is determined not simply by the frequency of co-occurrence between those two words, but by the total pattern of occurrence of all the words throughout the document. In most general terms, the synaptic weights depend not only on the co-occurrences of pairs of words, but by the co-occurrences in the context of all other occurrences.

Consequently, when a given term is activated, it will generate activations among all the other terms to which it is (substantially) connected. But when it is activated along with (in the context of) other terms, the set of other neurons that will be activated will be different.

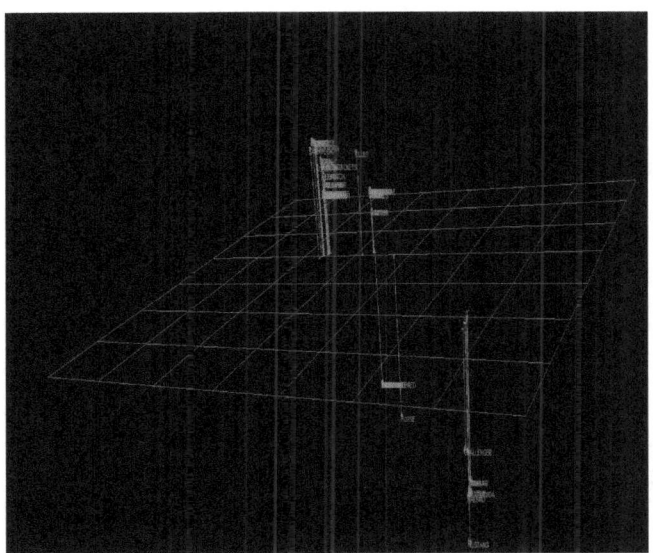

FIGURE 32: DISTANCES AMONG WORDS DERIVED FROM CATPAC SYNAPTIC WEIGHT MATRIX

Figure 32 provides an exact representation of the distances among items derived from the Catpac synaptic weights matrix by the Galileo[tm] algorithm[74] and displayed by ThoughtView[tr]. Unambiguous items that belong to one and only one category are located in identical places, so all their labels are printed on top of each other, which is why many of

115

the labels are illegible. Only ambiguous items, which occur in more than one category like *Mustang, Challenger, Firebird*, and others, or items that occur in more than one context (e.g., *horse*, which occurs in *Quarter Horse* and *Tennessee Walking Horse*, and *bee*, which occurs in *Killer Bee* and *Honey Bee*) or words that are associated with such words (e.g., *Honey* and *Killer*) exist outside of the well-defined categories, and hence are legible

Indstar

Indstar[tm] (**I**nteractive **N**eural **D**etection **S**torage and **R**etrieval) is a computer algorithm designed to extract this information from the synaptic weight matrix. Based on Catpac, Indstar replaces the parsing module of Catpac with an input module that reads lists like those in Figure 29. In this case, each of the 12 clusters is a list. A neural engine, identical to the one in Catpac III, then computes the synaptic weight matrix. Indstar replaces the third module of Catpac, the hierarchical cluster analysis algorithm, with a neural network designed to emulate the spatial network model documented by Dinauer and Fink[71]. This third neural module can be queried by activating any subset of its neurons, and Indstar will then report which other neurons are activated, along with their levels of activation. This is equivalent to asking "which other elements are members of a category that includes this subset of elements?", or, more generally, "which other elements are related to this subset of elements, and how related are they?"

The algorithm functions as follows: When one or more neurons are selected by the user, Indstar sets their activation value at 1.0 (activations are normalized to range between 0 and 1). Then all the remaining neurons in the system are polled. Each of them multiplies the activation value of the selected neurons by the synaptic weight connecting the selected neuron to the polled neuron and adds it to a temporary variable, called *anet*. Once the polled neuron has accumulated the net input to itself from all the selected neurons into *anet*, anet is then inserted into an *activation function* which calculates the activation value of that neuron. While the appropriate activation function remains an area of active research, Indstar offers a choice of four – a simple linear function, a logistic function ranging between zero and one, a logistic function ranging between ±1, and a hyperbolic tangent function that closely resembles a logistic function.

The default function in Indstar, and the one used in this research, is the logistic function ranging between ±1

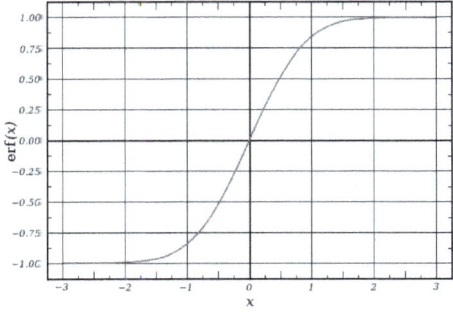

FIGURE 33: LOGISTIC FUNCTION VARYING BETWEEN -1 AND +1

While the exact shape of the logistic function varies depending on the values of the various coefficients in the equation, the general shape is found in virtually all organic functions, since, as Figure 33 shows, it represents no effect for stimuli too small to be detected by the organism, then rapid increase in effects once the threshold has been reached, followed by saturation where the organism cannot produce any more output no matter how high the input stimulus rises.

The use of the logistic function as a default also results from over a quarter century of research, beginning in the 1980's with research programs such as Johnson and Newton, then in the 1990's Catpac, Catpac 4 Windows and Oresme, in the 2000's with Catpac II, Wölfpak[75-77] and Listiac[66], and recently with Catpac III and Indstar.

```
    Enter as many items as you want, Your Uselessness. -1 when done)
MUSTANG
    Enter as many items as you want, Your Uselessness. -1 when done)
    -1
    And the winners are...


                          BARRACUDA                  Activation level = 0.0025
                          MUSTANG                     Activation level = 1.0000
                          CAMARO                      Activation level = 0.0025
                          FIREBIRD                    Activation level = 0.0025
                          COUGAR                      Activation level = 0.0025
                          JAVELIN                     Activation level = 0.0025
                          AMX                         Activation level = 0.0025
                          CHALLENGER                  Activation level = 0.0025
                          CLYDSDALE                   Activation level = 0.0025
                          ICELANDIC                   Activation level = 0.0025
                          PAINT                       Activation level = 0.0025
                          PALOMINO                    Activation level = 0.0025
                          PINTO                       Activation level = 0.0025
                          QUARTER HORSE               Activation level = 0.0025
                          SHETLAND PONY               Activation level = 0.0025
                          TENNESSEE WALKING HORSE     Activation level = 0.0025
                          THOROUGHBRED                Activation level = 0.0025
    One more time, Duke?
```

FIGURE 34: CONCEPTS ACTIVATED BY "MUSTANG."

118

Figure 34 shows that activating "Mustang" (activation set at 1.0) also activates all the elements of the car cluster and all of the elements of the horse cluster at about the same (.01) level. Changing the context, however, by activating both "Mustang" and "Camaro" changes the result, as Figure 35 shows:

```
 Enter as many items as you want, Your Uselessness. -1 when done)
 MUSTANG
  Enter as many items as you want, Your Uselessness. -1 when done)
 CAMARO
  Enter as many items as you want, Your Uselessness. -1 when done)
  -1
  And the winners are...


                    BARRACUDA                Activation level = 0.0050
                    MUSTANG                  Activation level = 1.0000
                    CAMARO                   Activation level = 1.0000
                    FIREBIRD                 Activation level = 0.0050
                    COUGAR                   Activation level = 0.0050
                    JAVELIN                  Activation level = 0.0050
                    AMX                      Activation level = 0.0050
                    CHALLENGER               Activation level = 0.0050
                    CLYDSDALE                Activation level = 0.0025
                    ICELANDIC                Activation level = 0.0025
                    PAINT                    Activation level = 0.0025
                    PALOMINO                 Activation level = 0.0025
                    PINTO                    Activation level = 0.0025
                    QUARTER HORSE            Activation level = 0.0025
                    SHETLAND PONY            Activation level = 0.0025
                    TENNESSEE WALKING HORSE  Activation level = 0.0025
                    THOROUGHBRED             Activation level = 0.0025
 One more time, Duke?
```

FIGURE 35: CONCEPTS ACTIVATED BY MUSTANG, CAMARO.

```
 Enter as many items as you want, Your Uselessness. -1 when done)
MUSTANG
 Enter as many items as you want, Your Uselessness. -1 when done)
CAMARO
 Enter as many items as you want, Your Uselessness. -1 when done)
AMX
 Enter as many items as you want, Your Uselessness. -1 when done)
-1
 And the winners are...


                    BARRACUDA                  Activation level = 0.0075
                    MUSTANG                    Activation level = 1.0000
                    CAMARO                     Activation level = 1.0000
                    FIREBIRD                   Activation level = 0.0075
                    COUGAR                     Activation level = 0.0075
                    JAVELIN                    Activation level = 0.0075
                    AMX                        Activation level = 1.0000
                    CHALLENGER                 Activation level = 0.0075
                    CLYDSDALE                  Activation level = 0.0025
                    ICELANDIC                  Activation level = 0.0025
                    PAINT                      Activation level = 0.0025
                    PALOMINO                   Activation level = 0.0025
                    PINTO                      Activation level = 0.0025
                    QUARTER HORSE              Activation level = 0.0025
                    SHETLAND PONY              Activation level = 0.0025
                    TENNESSEE WALKING HORSE    Activation level = 0.0025
                    THOROUGHBRED               Activation level = 0.0025
 One more time, Duke?
```

FIGURE 36: CONCEPTS ACTIVATED BY "MUSTANG", "CAMARO", "AMX"

```
 Enter as many items as you want, Your Uselessness. -1 when done)
PERRIER
 Enter as many items as you want, Your Uselessness. -1 when done)
-1
 And the winners are...

                    ENTERPRISE              Activation level = 0.0001
                    HORNET                  Activation level = 0.0001
                    WASP                    Activation level = 0.0001
                    INTREPID                Activation level = 0.0001
                    SARATOGA                Activation level = 0.0101
                    YORKTOWN                Activation level = 0.0001
                    MIDWAY                  Activation level = 0.0001
                    CORAL SEA               Activation level = 0.0001
                    EVIAN                   Activation level = 0.0100
                    PERRIER                 Activation level = 1.0000
                    PELLIGRINO              Activation level = 0.0100
                    FIJI                    Activation level = 0.0100
                    GLENLIVIT               Activation level = 0.0100
                    HIPPODROME              Activation level = 0.0001
                    BRIGHTON                Activation level = 0.0001
                    GALWAY                  Activation level = 0.0001
                    CHURCHILL DOWNS         Activation level = 0.0001
                    AQUADUCT                Activation level = 0.0001
 One more time, Duke?
```

FIGURE 37: CONCEPTS ACTIVATED BY "PERRIER"

As Figure 37 shows the effect of activating the unambiguous item *Perrier*, which belongs only to the sparkling water cluster. *Saratoga*, however, which also belongs to the sparkling water cluster, is an ambiguous concept that also belongs to the aircraft carrier cluster and the horse racing tracks cluster.

Activating an unambiguous item from a category that contains at least one ambiguous item will, of course, activate the other members of that category, but will also activate, at a much lower level, members of the other categories to which the ambiguous item belongs. This supports the results of Dinauer[70] and Dinauer and Fink[71]. Neither the aircraft carriers category nor the horse race tracks are subordinate categories to the sparkling water category, yet activation of a member of the sparkling water category activates elements of the aircraft carrier category and the horse racing tracks category because of the linkages among one of their constituent elements. Clearly, the spread of activation is not hierarchical.

Conclusions

The result provides support for the utility of the Indstar algorithm as a useful method of clustering data that is more consistent with the way human beings classify experience than conventional hierarchical cluster analysis. More importantly, results support earlier findings that show human cognitive processing more closely resembles a spatial network model than the traditional hierarchical categorical model.

Afterword: Galileo and Multidimensional Scaling

Since the early 1970's there has been some confusion about the relationship between Galileo and Multidimensional Scaling (MDS). Galileo, or The Galileo System as it is sometimes called, is the product mainly of sociologists and later communication scientists attempting to model cognitive processes as motions in space. Work began at the University of Wisconsin in Madison as part of a project initiated and overseen by Archie Haller designed to study the social psychological determinants of the educational and occupational aspirations of adolescent youth. The principle workers on the Galileo aspects of the project were Edward L. Fink and myself.

We had discovered that the educational and occupational aspirations of adolescent students were very well predicted by the average expectations of their significant others. The algebra that devolved from the average showed that the mean was the point of balance of forces (long known in physics, but new to us in sociology). Moreover, equations for attitude change represented a vector space. This implied

123

that attitude changes and other cognitive processes might be modeled as motions of points in a mathematical space.

Later, at the University of Illinois, I spent considerable time trying to find the coordinates of points in space from their interpoint distances, with no success. Statisticians at Illinois' SOUPAC (Statistically Oriented Users Package) suggested a variant of factor analysis as a method of determining the coordinates of points in such a space, and the first Galileo spaces were made using this model in Gail Wisan's dissertation (Wisan, 1969). She referred to the method as *unstandardized factor analysis*, which is a good description of the procedure. Later work, such as *"Standardized versus unstandardized data matrices: which type is best for factor analysis?"* (Woelfel, et. al., 1980) reprised this terminology.

An early study designed to test the precision of the Galileo procedures and their correspondence to conventional measurement systems used a questionnaire that referred to the units of measures as *galileos*[19] in recognition of Galileo's

----

[19] Specifically, the instructions said "Unlike physical distance, which is measured in feet or inches, psychological distance is measured in galileos." It then asked respondents to estimate how different or far apart a set of concepts were in the form "How different or far apart are (x) and (y) ? _____ gal." Since, for the N concepts, the question was repeated

use of comparative measurement procedures. Since the research design required respondents – Illinois sociology faculty and graduate students – to respond to the questionnaire three times, they began informally referring to the questionnaire and the technique as "galileo" – often in a phrase like, "Oh, no! Not another Galileo!"

When I moved to Michigan State University in 1973, SOUPAC was no longer available, and so a new method of computing the Galileo spatial coordinates was required. We were aware by then that a solution to the problem was known to some psychometricians and was referred to as *multidimensional scaling*. The problem was first raised – long before we were aware of it, and in fact before I was born -- by L. L. Thurstone, who referred to it as the "box problem."

Thurstone had spent many years trying to discover the factors underlying human intelligence, and had developed a technique known as *factor analysis* for this purpose. Although perhaps no other multivariate statistical technique ever experienced more revisions and variations, Thurstone was never satisfied that it accomplished his goal. When he retired, he left "the box problem" as a challenge to future scientists: if we have a series of boxes, and we make many measurements of them, (e.g., their width, height, surface area,

---

$N(N-1)/2$ times in each administration, respondents read the term "galileo" many times over the three administrations.

volume, etc.) can we discover that, underlying all of them, lie the dimensions *length, width* and *height*?

In 1938, Young and Householder in a classic article (Young and Householder, 1938) showed that the measurement required to find the dimensionality underlying a set of points in space was a matrix of their interpoint distances. The principle axes of the scalar products of the dissimilarities matrix were the dimensions sought in Thurstone's box problem.

Warren Torgerson modified this approach slightly by expressing all the vectors of the scalar products matrix from the center of the set of points by a procedure he called "double centering". The principle axes of this "double-centered scalar products" matrix were the dimensions of the space in which the points were projected with origin at the center of the set of points. Torgerson referred to this procedure as "Multidimensional Scaling."

Galileo researchers implemented the Young Householder Torgerson procedure for obtaining the double-centered scalar products matrix from a matrix of dissimilarities among the points in the Galileo algorithm. Kim Blaine Serota and Richard A. Holmes implemented the algorithm in FORTRAN, and used a method discovered by Karl Jacobi in 1849 and implemented in FORTRAN by Johannes Van de Geer in 1971 to calculate the eigenvalues and eigenvectors of this space. Although new to sociologists and Communication researchers, the mathematical procedure of projecting distances onto principle axes had been a core procedure in Physics for establishing inertial reference frames for well over a hundred years by this time.

While one of the central algorithms for multidimensional scaling became a part of the Galileo software, Galileo and multidimensional scaling parted ways at this point.

At the root of the divorce[20] were two core issues: first, Galileo theorists placed their primary trust in measurements, which they believed to mean exclusively comparison to some standard. MDS adherents, on the other hand, thought that the measurement of cognitive structures and processes was fundamentally impossible, and so whatever numbers resulted from measurements were considered untrustworthy.

Second, paired comparison data, particularly when well measured, seldom fit into low dimensional Euclidean spaces, but rather seemed to lie in higher dimensional Riemann spaces. Galileo researchers, committed to the data above all, took this as a finding, and began modifying the Galileo software to work in a general Riemannian space. , on the other hand, began devising mathematical algorithms that would modify observed values in such a way that the resulting

---

[20] It should be noted that, while Galileo researchers were very active in reading the MDS literature, attending MDS conferences and consulting with MDS experts, the corollary was not true, and MDS researchers had little or no interest in or knowledge of Galileo research.

modified dissimilarities would fit into low-dimensional Euclidean spaces. The basic goal of MDS researchers was to make low (preferably two) dimensional maps.

These so-called "non-metric" MDS algorithms, probably because they provided the hope that correct solutions could be found even though the original data were wrong, displayed a brief period of considerable popularity, particularly in market and advertising research, but ultimately proved less than useful and, several years later, some of the most prominent advocates of non-metric procedures conceded that the original Young Householder Torgerson techniques were generally more reliable[21].

---

[21] At the time the Galileo system was being developed, arguments in sociology and communication revolved around whether mathematical methods were appropriate for the study of human cognitive processes. A large part of *The Measurement of Communication Processes: Galileo Theory and Method* (Woelfel& Fink, 1980) was devoted to arguing for the utility of mathematics in studying human cognitive processes. It wasn't until many years of experience with psychometricians that I realized there were those who were enamored of mathematics to the exclusion of measurements. That same way of thinking is now prominent in network analysis.

Galileo researchers, on the other hand, moved toward increasing precision of measure, increasing study of the ways of tracking motions of points through the high dimensional Riemann space, and understanding the dynamics of such motions. In fact, on the flyleaf of their 1980 Galileo book, Woelfel & Fink say, "Most of the work done by communication researchers deals with how one operates on the multidimensional coordinate system yielded by the metric scaling algorithm rather than with how the coordinate system is generated."

This work included abstract theoretical work, such as study of how simple messages combine, how the self is defined across situations, modeling cultural processes as inertial processes in the reference frame, measuring cognitive aspects of the diffusion of innovations, measuring the inertial masses of concepts, the relationship of the network of concepts in the space to the networks of neurons in brains and culture, relationships between the meanings of texts and concept locations in the Galileo space, and many more abstract areas.

On the applied side, Galileo researchers use the system to model product market share, track election campaigns, design effective advertising and marketing strategies in many substantive areas across many disciplinary boundaries.

All through this period, Galileo researchers referred the system in divers ways, probably most often using the term "multidimensional scaling," or often "metric multidimensional scaling" to differentiate Galileo from the then more well-known non-metric models. In this they were adopting the psychometric usage, which, unfortunately, conflicts with established mathematical conventions. The "metric" algorithm, so called to distinguish it from the "non-metric" systems which treat measurements as if they are ordered values rather than magnitudes, is, in fact, not metric in the mathematical sense.

In mathematics, a metric space is one in which the triangle inequalities[22] are satisfied and is therefore Euclidean. This seldom turns out to be the case when the Young Householder Torgerson method is employed, since, empirically, the triangle inequalities rule is frequently violated in ways that make good sense and the resulting space is usually non-Euclidean, not metric.

Referring to the Galileo procedures as metric MDS, therefore, may make sense according to psychometric terminology, but it is clearly wrong in the much larger mathematical community.

---

[22] In any triangle drawn on a flat, Euclidean plane, any two sides must be at least as long as the third. If this is not true, the plane cannot be flat.

Even among psychometricians themselves, however, use of the term MDS to refer to Galileo is a cause of considerable confusion.

There is more to the meaning of words than their dictionary definition. In the 1980's, a joint conference of Galileo researchers and prominent psychometricians, including Robert Pruzek and James Ramsay (at that time president of the Psychometric Society) was held at the University at Albany. At first, the two groups had considerable difficulty understanding each other, until Ramsay explained that, to the psychometricians, "MDS" referred to a specific set of people who did specific things, but what the Galileo researchers was doing was not related to their work. He suggested we don't use the term "multidimensional scaling", but instead refer to ourselves by the name people associate with us -- *Galileo*. There s no need to identify Galileo work with a different research thread from a different discipline. It only causes confusion. Better to say that Galileo, while it shares some ancestral roots with multidimensional scaling, goes far beyond MDS into the analysis of cognitive and cultural processes.

References

1.   G. Wisan, Doctoral dissertation, University of Illinois at Urbana-Champaign, 1972.
2.   J. Woelfel *The Culture of Science: Is Social Science Science?* (Rah Press, Buffalo, NY, 2016).
3.   E. Durkheim, *The Rules of Sociological Method (1895), 8th edition, trans. Sarah A. Solovay and John M. Mueller, ed. George E. G. Catlin (1938, 1964 edition), 13.,* 8, 1934, 1968 ed. (1895).
4.   M. Jahoda, *Research Methods in Social Relations: Selected techniques.* (Dryden Press, 1951).
5.   W. H. Sewell, Social Psychology Quarterly **52** (Jun, 1989), 11 (1989).
6.   J. Platt, *A History of Sociological Research Methods in America, 1920-1960.* (Cambridge University Press, Cambridge, 1998).
7.   E. Schrödinger, *Nature and the Greeks and Science and Humanism.* (Cambridge University Press, Cambridge, 2002).
8.   P. Berger and T. Luckman, *The Social Construction of Reality.* (Doubeday, Garden City, NJ, 1966).
9.   E. Iacobucci, University at Buffalo, State University of New York, 2016.

10.   E. Larson, R. Darilek, D. Gibran, B. Nichiporuk, A. Richardson, L. Schwartz and C. Thurston, *Foundations of effective influence operations: A framework for enhancing army capabilities*. (RAND Corporation. Available online at http://www.rand.org/pubs/monographs/2009/RAND_MG654.pdf, Santa Monica, CA, 2009).

11.   E. Iacobucci, University at Buffalo, State University of New York, 2016.

12.   J. Woelfel and N. Stoyanoff, in *The role of communication in business transactions and relationships*, edited by M. Hinner (Peter Lang, Berlin, Germany, 2007), Vol. 3, pp. 433-462.

13.   R. McIntosh and J. Woelfel Communication & Science Journal (2018).

14.   G. A. Barnett, in *Readings in the Galileo system: Theory, methods, and application*, edited by G. A. Barnett and J. Woelfel (Kendall/Hunt, Dubuque, IA, 1988), pp. 243-264.

15.   G. A. Barnett, K. Serota and J. Taylor, Human Communication Research **2** (3), 227-244 (1976).

16.   J. Woelfel, E. L. Fink, R. Holmes, M. Cody and J. Taylor, (Retrieved from Galileoco, Galileo Literature #40. (http://www.galileoco.com/CEtestLit/literature.asp), 1974).

17.   D. Kincaid, J. Yum, J. Woelfel and G. A. Barnett, Quality and Quantity **18**, 59-79 (1983).

18.   B. Newton, E. Buck and J. Woelfel, Human Organization **45** (2), 162-170 (1986).

19.   P. Cheong, J. Hwang, B. Elbirt, H. Chen, C. Evans and J. Woelfel, in *The interrelationship of business and communication*, edited by M. Hinner (Peter Lang, Berlin, 2010), Vol. 6.

20.   J. Woelfel, in *Career behavior of special groups*, edited by J. S. Picou and R. E. Campbell (Merrill, Columbus, OH, 1975), pp. 41-61.

21.   J. Saltiel, Quality & Quantity **24**, 283-296 (1990).

22.   J. Saltiel, Work and Occupations **15** (3), 334-355 (1988).

23.   J. Saltiel, in *Readings in the Galileo system: Theory, methods and applications*, edited by G. A. Barnett and J. Woelfel (Kendall/Hunt, Dubuque, IA, 1988), pp. 295-312.

24.   J. Saltiel, *An application of a social psychological model to the problem of occupational choice.* (RAH Press, Amherst, NY, 2009).

25.   J. Saltiel, in *First Annual Conference on Metric Multidimensional Scaling, International Communication Association* (Chicago, 1978).

26.   G. A. Barnett, M. Palmer and H. Al-Deen, in *Communication yearbook 8*, edited by R. Bostrom (Sage Publications, Beverly Hills, CA, 1984), pp. 659-677.

27. K. H. Kwon, G. A. Barnett and H. Chen, Journal of International and Intercultural Communication **2** (2), 107-138 (2009).

28. D. Brandt and G. A. Barnett, (Retrieved from Galileoco, Galileo Literature #36. (http://www.galileoco.com/CEtestLit/literature.asp), 1979).

29. P. Caran and M. Hennessey, Sociological Perspectives **32** (2), 227-243 (1989).

30. F. Korzenny, N. J. Stoyanoff, M. Ruiz and A. David, International Journal of Intercultural Relations **4**, 77-95 (1980).

31. G. A. Barnett, R. Wigand, R. Harrison, J. Woelfel and A. Cohen, Human Organization **40** (4), 330-337 (1981).

32. K. Serota, E. L Fink, J. Noell and J. Woelfel, in *April 1975 International Communication Association conference* (Chicago, IL, 1976).

33. G. A. Barnett, in *workshop on metric multidimensional scaling, International Communication Association conference* (Philadelphia, PA, 1979).

34. J. Woelfel in *Social Science and Social Computing: Steps to Integration* (University of Hawaii under a grant from the Air Force Office of Scientific Research, Honolulu, HI, 2010).

35. J. Woelfel and M. Murero, Research in Social Stratification and Mobility **22**, 51-71 (2005).

36.   P. Barresi, J. Berens, S. Henne, C. Iacobucci and E. Redmond, (Retrieved from Galileoco, Galileo Literature #44. (http://www.galileoco.com/CEtestLit/literature.asp), 2000).

37.   C. Colfer, Y. Byron, R. Prabhu and E. Wollenberg, (n.d.).

38.   C. J. Colfer, J. Woelfel, R. Wadley and E. Harwell, in *People managing forests: The links between human well-being and sustainability*, edited by C. J. P. Colfer and Y. Byron (Resources for the Future, Washington, D.C., 2001), pp. 135-154.

39.   S. Allen, The Forestry Chronicle **81** (1), 381-386 (2005).

40.   E. L. Fink, J. Robinson and S. Dowden, Communication Research **12** (3), 301-318 (1985).

41.   E. L. Fink and S. Chen, Human Communication Research **21**, 494-521 (1995).

42.   H. C. DeLeo, MA thesis, Temple University, Philadelphia, PA., 1976.

43.   J. Foldy and J. Woelfel, Quality & Quantity **24**, 1-16 (1990).

44.   J. Cary, A. Beel and H. Hawkins, *Farmers' attitudes towards land management for conservation.* (FD Atkinson Government Printer, School of Agriculture and Forestry, University of Melbourne. Melbourne, Australia, 1986).

45.   R. Wilkinson and J. Cary, *Monitoring landcare in central Victoria.* (School of Agriculture and Forestry, University of Melbourne, 1992).
46.   A. Vishwanath and H. Chen, Journal of the American Society for Information Science and Technology **57** (11), 1451-1460 (2006).
47.   R. Hsieh, Doctoral dissertation, University at Buffalo, 2004.
48.   G. Fitzgerald, L. Saunders and R. Wilkinson, (1995).
49.   J. O. Yum, The Journal of Social Psychology **128** (6), 765-777 (1988).
50.   J. Woelfel, in *Progress in communication sciences 13*, edited by G. A. Barnett and F. Boster (Norwood, NJ: Ablex Publishing, 1997), pp. 213-227.
51.   J. Woelfel, M. J. Cody, J. Gillham and R. A. Holmes, Human Communication Research **6** (2), 153-167 (1980).
52.   Y. Lim, University at Buffalo, State University of New York, 2008.
53.   K. Rezaei-Moghaddam, E. Karami and J. Woelfel, Journal of Food, Agriculture & Environment **4** (2), 310-319 (2006).
54.   H. Marshall, Doctoral dissertation, University at Buffalo, 2006.
55.   S. Bass, T. Gordon, S. Ruzek and A. Hausman, Biosecurity and Bioterrorism: Biodefense Strategy, Practice, and Science **6** (2), 179-190 (2008).

56.	P. Studtmann, in *The Stanford Encyclopedia of Philosophy*, edited by E. N. Zalta (Summer, 2013).
57.	D. Elgin and A. Mitchell, Co-Evolution Quarterly (Summer) (1977).
58.	N. Chomsky, *Syntactic Structures*. (Mouton, The Hague, 1957).
59.	J. R. Searle, The New Yor Review of Books **18** (112) (1972).
60.	J. A. Danowski, in *Progress in Communication Sciences XII*, edited by G. A. Barnett and W. Richards (Ablex, NJ, 1993), pp. 197-222.
61.	J. Woelfel, *Artificial neural networks for cluster analysis*. (RAH Press, Amherst, NY, 2009).
62.	J. Woelfel, Journal of Communication **43** (1), 63-80 (1993).
63.	J. A. Danowski, in *Communication Yearbook 5*, edited by M. Burgoon (Transaction Books, New Nrunswick, N.J., 1982).
64.	J. A. Danowski, TREC, 5 (1992).
65.	R. S. Porter, MD and J. L. Kaplan, MD (Eds), (Merck Sharp & Dohme Corp.,

a subsidiary of Merck & Co., Inc, Whitehouse Station, N.J., U.S.A., 2004-2012).
66.	B. Battleson, H. Chen, C. Evans and J. Woelfel in *Sunbelt XXVII, INSNA Social Networking Conference* (St. Pete Beach, FL, 2008).
67.	C. Spearman, American Journal of Psychology **15**, 92 (1904).

68.   F. Murtagh and M. J. Kurtz, (Cornell University, Cornell Univesity Library Archives, 2013).
69.   M. El-Melegy, E. A. Zanaty, W. M. Abd-Elhafiez and A. Faraq, presented at the Image Proessing, 2007. ICIP 2007. IEEE International Conference on, Vol. 6:, San Antonio, TX USA. 2007 (unpublished).
70.   L. Dinauer, Doctoral dissertation, University of Maryland, 2003.
71.   L. Dinauer and E. L. Fink, Human Communication Research **31** (1), 1-32 (2005).
72.   C. Evans, (2013), pp. Catpac Bibliography.
73.   S. C. Johnson, Psychometrika **2**, 241-254 (1967).
74.   J. Woelfel and E. L. Fink, *The measurement of communication processes: Galileo theory and method.* (Academic Press, New York, 1980).
75.   C. C. Evans, Hao; Battleson, Brenda; Wölfel, Joseph K.; Woelfel, Joseph, RAH Press, 27 (2010).
76.   J. Wölfel, H. Chen, J. Kim, M. Murero, B. Sharma, J. Woelfel and R. Hsieh, in *International Communication Association conference* (New York, NY, 2005, Oct.).
77.   J. Wölfel, R. Hsieh, H. Chen, J. Hwang, P. Cheong, D. Rosen and J. Woelfel, in *25th Annual Meeting of the International Network for Social Network Analysis (INSNA) conference* (Redondo Beach, CA, 2005, February).

Index

U

Univac, 37
University of Illinois, 125

V

VALS, 80, 81

W

WH3, 60, 61, 63
Wisconsin Significant Other Project, 25

Y

Yourself, 32

www.ingramcontent.com/pod-product-compliance
Lightning Source LLC
Chambersburg PA
CBHW070252190526
45169CB00001B/375